ROBUST CONTROL SYSTEMS with GENETIC ALGORITHMS

CONTROL SERIES

Robert H. Bishop
Series Editor
University of Texas at Austin
Austin, Texas

Published Titles

Linear Systems Properties: A Quick Reference
Venkatarama Krishnan

Robust Control Systems and Genetic Algorithms
*Mo Jamshidi, Renato A Krohling, Leandro dos Santos Coelho,
and Peter J. Fleming*

Sensitivity of Automatic Control Systems
Efim Rozenwasser and Rafael Yusupov

Forthcoming Titles

Material and Device Characterization Measurements
Lev I. Berger

Model-Based Predictive Control: A Practical Approach
J.A. Rossiter

ROBUST CONTROL SYSTEMS with GENETIC ALGORITHMS

Mo Jamshidi
Renato A. Krohling
Leandro dos Santos Coelho
Peter J. Fleming

CRC Press
Taylor & Francis Group
Boca Raton London New York

CRC Press is an imprint of the
Taylor & Francis Group, an **informa** business

CRC Press
Taylor & Francis Group
6000 Broken Sound Parkway NW, Suite 300
Boca Raton, FL 33487-2742

First issued in paperback 2019

© 2003 by Taylor & Francis Group, LLC
CRC Press is an imprint of Taylor & Francis Group, an Informa business

No claim to original U.S. Government works

ISBN-13: 978-0-8493-1251-9 (hbk)
ISBN-13: 978-0-367-39572-8 (pbk)

Library of Congress Cataloging-in-Publication Data

Robust control systems with genetic algorithms / Mohammad Jamshidi ... [et al.].
    p. cm. -- (Control series)
Includes bibliographical references and index.
ISBN 0-8493-1251-5 (alk. paper)
    1. Robust control. 2. I. Jamshidi, Mohammad. II. CRC Press control series.

TJ217.2 .R62 2002
629.8--dc21

2002073574

Library of Congress Card Number 2002073574

Visit the Taylor & Francis Web site at
http://www.taylorandfrancis.com

and the CRC Press Web site at
http://www.crcpress.com

Dedications

In memory of my fathers: Habib Jamshidi and General Morteza Salari
Mo Jamshidi

For my mother Hilda Stern Krohling
and in memory of my father Daniel Krohling
Renato Krohling

For Viviana
Leandro dos Santos Coelho

For Steph, Joss, and Sam
Peter J. Fleming

Preface

Since the early days of the latter part of the last century, optimization has been an integral part of many science and engineering fields such as mathematics, operations research, control systems, etc. A number of approaches have existed to bring about optimal behavior in a process of plant. Traditionally, mathematics is the basis for many optimization approaches, such as optimal control with such celebrated theoretical results such as Pontryagin's maximum principle, Hamilton–Jacobi–Bellman sufficiency equation, Kuhn–Tucker conditions, etc. In recent times, we have witnessed a new paradigm for optimization — a biologically inspired approach has arrived that is based on the natural evolution of populations to Darwin's principle of natural selection, "survival of the fittest," and Mendel's genetics law of transfer of the hereditary factors from parents to descendants.

The principal player in this evolutionary approach to optimization is known as genetic algorithms (GA), which was developed by Holland in 1975, and is based on the principles of natural selection and genetic modification. GA are optimization methods, which operate on a population of points, designated as individuals. Each individual of the population represents a possible solution of the optimization problem. Individuals are evaluated depending upon their fitness. The fitness indicates how well an individual of the population solves the optimization problem. Another paradigm is based on optimization of symbolic codes, such as expert rules of inference, and is known as genetic programming (GP), first suggested by Koza in 1992. GP is an extension of the GA for handling complex computational structures. The GP uses a different individual representation and genetic operators and an efficient data structure for the generation of symbolic expressions, and it performs symbolic regressions. The solution of a problem by means of GP can be considered a search through combinations of symbolic expressions. Each expression is coded in a tree structure, also called a computational program, that is subdivided into nodes and presents a variable size.

One of the popular approaches to the mathematics-based approach to optimal design of a control system has been robust optimal control, in which an objective function, often based on a norm of a functional, is optimized, while a controller (dynamic or static) is obtained that can tolerate variation of plant parameters and unordered dynamics.

The object of this volume is to build a bridge between genetic algorithms and robust control design of control systems. GA is used to find an optimal robust controller for linear control systems. Optimal control of robotic manipulators, flexible links, and jet engines are among the applications considered in this book.

In Chapter 1, an introduction to genetic algorithms is given, showing the basic elements of this biologically inspired stochastic parallel optimization method. The chapter describes the genetic operators: selection, cross-over, and mutation, for binary and real representations. An example of how genetic algorithms work as an optimizer is provided, followed by a short overview of genetic programming.

Chapter 2 is devoted to optimal design of robust control systems and addresses issues like robust stability and disturbance rejection. First, by means of the H_∞- norm, two conditions are described, one for robust stability and one for disturbance rejection. Finally, the design of optimal robust controllers and the design of optimal disturbance rejection controllers, both with fixed structure, are formulated as a constrained optimization problem. The problem consists of the minimization of a performance index (the integral of the squared-error or the integral of the time-weighted-squared-error) subject to the robust stability constraint or the disturbance rejection constraint, respectively. The controller design, therefore, consists in the solution of the constrained optimization problems.

Chapter 3 is concerned with new methods to solve these optimization problems using genetic algorithms. The solution contains two genetic algorithms. One genetic algorithm minimizes the performance index (the integral of the squared-error or the integral of the time-weighted-squared-error) and the other maximizes the robust stability constraint or the disturbance rejection constraint. The entire design process is described in the form of algorithms. The methods for controller design are evaluated, and the advantages are highlighted.

Chapter 4 deals with model-based predictive control and variable structure control designs. The basic concepts and formulation of generalized predictive control based on optimization by genetic algorithms are presented and discussed. The variable structure control design with genetic optimization for control of discrete-time systems is also presented in this chapter.

Chapter 5 is devoted to the development of a genetic algorithm to the design of generalized predictive control and variable structure systems of quasi-sliding mode type. The simulation results for case studies show the effectiveness of the proposed control schemes.

Chapter 6 discusses the use of fuzzy logic in controllers and describes the role of genetic algorithms for off-line tuning of fuzzy controllers. As an example application, a fuzzy controller is developed and tuned for a gas turbine engine.

Chapters 7 and 8 take on the application of hybrid approaches such as GA-Fuzzy and Fuzzy-GP to robotic manipulators and mobile robots with some potential applications for space exploration. The fuzzy behavior control approach with GP enhancement or fuzzy control optimized by a GA is among the key topics in these two chapters.

Chapter 9 addresses the simultaneous optimization of multiple competing design objectives in control system design. A multiobjective genetic algorithm is introduced, and its application to the design of a robust controller for a benchmark industrial problem is described.

In Appendix A, we cover the fundamental concepts of fuzzy sets, fuzzy relation, fuzzy logic, fuzzy control, fuzzification, defuzzification, and stability of fuzzy control systems. This appendix is provided to give background to the readers who may not be familiar with fuzzy systems, and it is a recommended reading before Chapters 6 through 8.

Mo Jamshidi would like to take this opportunity to thank many peers and colleagues and current and former students. In particular, much appreciation is due to two of the former doctoral students of the first author, Dr. Edward Tunstel of NASA Jet Propulsion Laboratory (Pasadena, California) whose fuzzy-behavior control approach was the basis for Chapter 8, as well as Dr. Mohammad Akbarzadeh-Totoonchi of Ferdowsi University (Mashad, Iran), whose interests in GA always inspired the first author and whose work on GA-Fuzzy control of distributed parameter systems made Chapter 7 possible. He wishes to express his heart-felt appreciation to his family, especially his loving wife Jila, for inspiration and constant support.

Renato A. Krohling wishes to kindly thank his family for their constant support and their motivation throughout his career. He wishes to thank LAI-UFES, the Intelligent Automation Laboratory, Electrical Engineering Department, UFES, Vitória-ES, Brazil, for their support of his work. He would like to especially thank the guidance and mentorship of Professor H.J.A. Schneebeli, his MS degree advisor, and the coordinators of the "Programa de Pós-Graduação em Engeharia Elétrica, UFES." A great part of the research of Chapters 1 to 3 was prepared when he was a doctoral student at the University of Saarland, Germany, and he appreciates the guidance of his advisor Professor H. Jaschek. In the last few years, he has also carried out some works with Professor J.P. Rey, NHL, Leeuwarden, the Netherlands, and thanks him for very useful "scientific hints." He wishes to thank the Brazilian Research Council, "Conselho Nacional de Pesquisa e Desenvolvimento Científico e Tecnológico (CNPq)," for their financial support. He wishes to thank his nephew Helder for art works, his brothers and sisters of KAFFEE Exp. & Imp. Ltd., Santa Maria, ES, for their financial support, allowing his stay in the United States, hosted by first author, Mo Jamshidi, Director of the ACE Center, the University of New Mexico, during the summer of 2000.

Leandro dos S. Coelho wishes to thank his wonderful wife and his family for their love, support and encouragement. He wishes to thank "Programa de Pós-Graduação em Engenharia de Produção e Sistemas, Pontifícia Universidade Católica do Paraná," for their support of his work. He would like to especially thank Professor Antonio Augusto Rodrigues Coelho, his doctoral degree advisor, at Federal University of Santa Catarina, Brazil. He also wishes to thank the Brazilian Research Council, "Conselho Nacional de

Pesquisa e Desenvolvimento Científico e Tecnológico (CNPq)," for the financial support during his doctoral studies.

Peter Fleming is indebted to two of his former doctoral students, Beatrice Bica and Ian Griffin, for permission to use some of their excellent research results in Chapters 6 and 9, respectively. He also wishes to take this opportunity to thank all the researchers who have contributed to make the Badger Lane research laboratory such a vibrant and productive environment during the last ten years.

The encouragement and patience of CRC Press LLC Editor Nora Konopka is very much appreciated. Without her continuous help and assistance during the entire course of this project, we could not have accomplished the task of integrating GA and robust control under the cover of this volume. We would also like to thank the patient collaboration and helpful assistance of Jamie Sigal and Joette Lynch of CRC Press through the production phase of the book and their commitment and skillful effort of editing and processing several iterations of the manuscript. Finally, we would like to thank Dr. Robert H. Bishop, the Series Editor of this volume for the valuable reviews of the manuscripts and helpful suggestions during the initial stages of the book. Last, but not least, the authors would like to thank their families for their understanding, love, and patience during this project.

Mo Jamshidi
Albuquerque, New Mexico, United States

Renato A. Krohling
Vitória Espírito Santo, Brazil

Leandro dos S. Coelho
Curitiba, Brazil

Peter Fleming
Sheffield, United Kingdom

Table of Contents

chapter one

Genetic algorithms

1.1 Introduction to genetic algorithms

The evolutionary theory attributes the process of the natural evolution of populations to Darwin's principle of natural selection, "survival of the fittest," and Mendel's genetics law of transfer of the hereditary factors from parents to descendants (Michalewicz, 1996). Genetic algorithms (GA) were developed by Holland (1975) and are based on the principles of natural selection and genetic modification. GA are optimization methods, which operate on a population of points, designated as individuals. Each individual of the population represents a possible solution of the optimization problem. Individuals are evaluated depending upon their fitness. Fitness indicates how well an individual of the population solves the optimization problem.

GA begin with random initialization of the population. The transition of one population to the next takes place via the application of the genetic operators: *selection, crossover,* and *mutation.* Through the selection process, the fittest individuals will be chosen to go to the next population. Crossover exchanges the genetic material of two individuals, creating two new individuals. Mutation arbitrarily changes the genetic material of an individual. The application of the genetic operators upon the individuals of the population continues until a sufficiently good solution of the optimization problem is found. The solution is usually achieved when a predefined stop condition, i.e., a certain number of generations, is reached. GA have the following general features:

- GA operate with a population of possible solutions (individuals) instead of a single individual. Thus, the search is carried out in a parallel form.
- GA are able to find optimal or suboptimal solutions in complex and large search spaces. Moreover, GA are applicable to nonlinear optimization problems with constraints that can be defined in discrete or continuous search spaces.

- GA examine many possible solutions at the same time. So, there is a higher probability that the search converges to an optimal solution.

In the classical GA developed by Holland (1975), individuals are represented by binary numbers, i.e., bit strings. In the meantime, new representations for individuals and appropriate genetic operators have been developed. For optimization problems with variables within the continuous domain, the real representation has shown to be more suitable. With this type of representation, individuals are represented directly as real numbers. For this case, it is not necessary to transform real numbers into binary numbers. In the following, some terms and definitions are described.

1.2   Terms and definitions

Here is an introduction to common GA terms. Because GA try to imitate the process of natural evolution, the terminology is similar but not identical to the terms in natural genetics. A detailed description of the GA can be found in Goldberg (1989), Michalewicz (1996), Mitchell (1996), Bäck, (1996), and Fogel (1995).

For GA, population is a central term. A population P consists of individuals c_i with $i = 1,\dots,\mu$:

$$P = \left\{ c_1, \dots, c_i, \dots, c_\mu \right\} \tag{1.1}$$

The population size μ can be modified during the optimization process. In this work, however, it is kept constant.

An individual is a possible solution of an optimization problem. The objective function $f(x)$ of the optimization problem is a scalar-valued function of an n-dimensional vector x. The vector x consists of n variables x_j, with $j = 1,\dots,n$, which represents a point in real space $\Re^n$. The variables x_j are called genes. Thus, an individual c_i consists of n genes:

$$c_i = \left[c_{i1}, \dots, c_{ij}, \dots, c_{in} \right] \tag{1.2}$$

In the original formulation of GA, the individuals were represented as binary numbers consisting of bits 0 and 1. In this case, the binary coding and the Gray coding can be used. An individual c_i coded in binary form is called a *chromosome*. In real representation, an individual c_i consists of a vector of real numbers x_i.

Another important term is the fitness of an individual. It is the measure for the quality of an individual in a population. The fitness of an individual is determined by the fitness function $F(x)$. GA always search for the fittest individual, so the fitness function is maximized. In the simplest cases, the fitness function depends on the objective function.

The average fitness F_m of a population is determined as follows:

$$F_m = \frac{\sum\limits_{i=1}^{\mu} F(\mathbf{c}_i)}{\mu} \tag{1.3}$$

The relative fitness p_i of an individual c_i is calculated by:

$$p_i = \frac{F(\mathbf{c}_i)}{\sum\limits_{i=1}^{\mu} F(\mathbf{c}_i)} \tag{1.4}$$

Normally, GA begin the process of optimization with a randomly selected population of individuals. Then, the fitness for each individual is calculated. Next comes the application of the genetic operators: *selection, crossover,* and *mutation.* Thus, new individuals are produced from this process, which then form the next population. The transition of a population P_g to the next population P_{g+1} is called generation, where g designates the generation number. In Figure 1.1, the operations executed during a generation are schematically represented. The evolution of the population continues through several generations, until the problem is solved, which in most cases, ends in a maximum number of generations g_{max}.

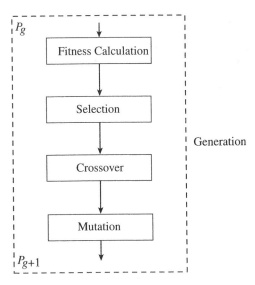

Figure 1.1 Representation of the executed operations during a generation.

1.3   Representation

In the classical GA, the genetic operators crossover and mutation operate on bit strings. Therefore, the fitness of individuals can only be calculated after decoding the bit strings. The next section describes GA with binary and real representation.

1.3.1   Genetic algorithms with binary representation

Binary representation consists of binary coding and Gray coding. In order to present the principles of the classical GA, only binary coding is described. For a description of Gray coding, the reader is referred to Bethke (1981).

Let the objective function be $f(x)$, where the vector x consists of n variable x_i with $i = 1,...,$n. The lower and upper value of the variables x_i is, respectively, given by $x_{i\,min}$ and $x_{i\,max}$. In binary coding, the variable x_i is first converted into a normalized number $x_{i\,norm}$:

$$x_{i\,norm} = \frac{x_i - x_{i\,min}}{x_{i\,max} - x_{i\,min}} \tag{1.5}$$

where $0 \leq x_{i\,norm} \leq 1$. Next, the normalized number $x_{i\,norm}$ is transformed into a binary number c_i. The number of bits required for c_i is determined by the accuracy required. The binary number c_i consists, thus, of m bits and is represented as follows:

$$c_i = \langle b[1],...,b[j],...,b[m] \rangle \qquad \text{with} \quad b[j] \in \{0,1\} \qquad \forall j = 1,...,m \tag{1.6}$$

The encoding of a normalized number $x_{i\,norm}$ into the corresponding binary number c_i occurs in accordance with the pseudo-code shown in Figure 1.2.

The decoding of a binary number c_i into the correspondent variable x_i occurs in accordance with the pseudo-code shown in Figure 1.3.

1.3.2   Genetic algorithms with real representation

For optimization problems with variables in continuous domain, a representation with real numbers is easier and more direct. An individual c_i consists of a vector of real numbers x_i. Each element of the vector corresponds to a characteristic of the individual, thus, a gene. Therefore, no coding or decoding is needed. This leads to a simpler and more efficient implementation. The accuracy with real representation depends only on the computer used.

The advantages of the real representation compared to the binary representation are shown in Davis (1991), Wright (1991), and Michalewicz (1996).

Algorithm 1: binary coding

Input: $x_{i\,\text{norm}}$ and m

Output: $c_i = \langle b[1], \ldots, b[j], \ldots, b[m] \rangle$

Auxiliary variable: j and q_j

begin

$\quad\quad j = 1$ and $q_j = x_{i\,\text{norm}}$

$\quad\quad$ **while** $\left(j \leq m \right)$ **do**

$\quad\quad\quad$ **if** $\left(q_j - 2^{-j} \geq 0 \right)$

$\quad\quad\quad\quad b[j] = 1$

$\quad\quad\quad\quad q_{j+1} = q_j - 2^{-j}$

$\quad\quad\quad$ **else**

$\quad\quad\quad\quad b[j] = 0$

$\quad\quad\quad\quad q_{j+1} = q_j$

$\quad\quad\quad$ **end if**

$\quad\quad\quad j = j + 1$

$\quad\quad$ **end while**

$\quad\quad$ **return** $c_i = \langle b[1], \ldots, b[j], \ldots, b[m] \rangle$

end

Figure 1.2 Pseudo-code for the binary coding.

1.4   Fitness function

In GA, each individual represents one point in the search space. The individuals undergo a simulated evolutionary process. In each generation, relatively "good individuals" reproduce, while relatively "bad individuals" do not survive. The fitness of an individual serves to differentiate between relatively good and bad individuals. The fitness of an individual is calculated by the fitness function $F(x)$. For optimization problems without constraints, the fitness function depends on the objective function of the optimization problem on hand. For maximization problems, the fitness function is calculated by the following (Goldberg, 1989):

$$F(x) = \begin{cases} f(x) + C_{\min}, & \text{if } f(x) + C_{\min} > 0 \\ 0, & \text{otherwise} \end{cases} \tag{1.7}$$

The constant C_{min} is a lower bound for the fitness. The introduction of the constant C_{min} is necessary for the mapping of a negative fitness into a positive fitness, because many selection methods require a nonnegative fitness.

Algorithm 2: binary *decoding*

Input: $c_i = \langle b[1], \ldots, b[j], \ldots, b[m] \rangle$

Output: $x_{i\,\text{norm}}$

Auxiliary variable: j and q_j

begin

    $j = 1$

    $q_j = 0$

    while $(j \leq m)$ **do**

    if $(b[j] = 0)$

        $q_{j=1} = q_j$

    else

        $q_{j+1 q_j} = q_j + 2^{-j}$

    end if

    $j = j + 1$

    end while

    $x_{i\,\text{norm}} = q_j + 1$

    return $x_{i\,\text{norm}}$

end

Figure 1.3 Pseudo-code for binary decodification.

For minimization problems, it is necessary to change the objective function, because GA work according to the principle of the maximization of fitness. By multiplying the objective function with the factor of minus one (–1), the minimization problem is transformed into a maximization problem. The fitness function is then calculated by the following (Goldberg, 1989):

$$F(x) = \begin{cases} C_{\text{max}} - f(x), & \text{if } C_{\text{max}} > f(x) \\ 0, & \text{otherwise} \end{cases} \tag{1.8}$$

The constant C_{max} is an upper bound for the fitness and is introduced here to map negative fitness into positive fitness. Details on how to determine the constants C_{min} and C_{max} or other methods for fitness scaling are not given here, because in this work, we use the tournament selection method, which does not require positive fitness, i.e., negative fitness is also allowed. Therefore, C_{min} for maximization problems in Equation (1.7) can be set to zero. Thus, it follows that:

$$F(\mathbf{x}) = f(\mathbf{x}) \tag{1.9}$$

For minimization problems, C_{max} can be set to zero in Equation (1.8), resulting in

$$F(\mathbf{x}) = -f(\mathbf{x}) \tag{1.10}$$

For constraint optimization problems, the fitness function becomes more complex (see next chapter).

1.5   Genetic operators

Through the application of the genetic operators *selection, crossover,* and *mutation*, GA generate a new population from an existing population. In the following section, these three operators are described.

1.5.1   Selection

The *selection* process chooses the fittest individuals from a population to continue into the next generation. GA use Darwin's principle of natural selection, "survival of the fittest," to select individuals. *Selection* compares the fitness of one individual in relation to other individuals and decides which individual goes on to the next population. Through selection, "good individuals" are favored to advance with high probability, while "bad individuals" advance with low probability to the next generation.

Here, another important term, *selection pressure*, is introduced. Selection pressure is the degree to which the better (fitter) individuals are favored: the higher the selection pressure, the more better individuals are favored (Miller and Goldberg, 1995). A selection pressure in the GA that is too high might cause premature convergence to a local optimum. Conversely, a selection pressure that is too low can lead to a slow convergence. The convergence rate of the GA is determined to a wide extent by the selection pressure. The GA is able to find optimal or suboptimal solutions under different selection pressures (Goldberg et al., 1993).

Theoretical investigations and comparisons of the efficiency of different selection methods can be found in Blickle (1997). In the following section, two of these are described: *proportionate selection*, which was used with the classical GA, and *tournament selection*, which will be used in this work.

1.5.1.1   Proportionate Selection

In proportionate selection, the probability of selection, i.e., the probability that an individual c_i advances to the next generation, is proportionate to its relative fitness p_i. The expected number ξ of offspring of an individual c_i is obtained by the product of the relative fitness p_i times the number of individuals of the population μ, i.e., $\xi = p_i \cdot \mu$. Proportionate selection can be illustrated by using a roulette wheel as an example. The number of fields of the roulette wheel correspond to the population size. Each individual is assigned exactly one field on the roulette wheel. The size of the field is proportional to the fitness of the individual belonging to the field. The probability that the marker stops on a certain field is equal to the relative fitness of that individual. The roulette is spun μ times, which corresponds to the population size.

Proportionate selection, developed originally by Holland for classical GA, is only applicable to nonnegative fitness. For negative fitness, it is necessary to use a *fitness scaling* method (Goldberg, 1989). Studies indicate that the tournament selection method presents better performance (Blickle, 1997).

1.5.1.2   Tournament selection

Here, the fittest individual chosen from a group of z individuals of the population advances to the next population. This process is repeated μ times. The size of the group of z is called *tournament size*. The selection pressure can be easily increased by increasing the tournament size. On average, the winner of a larger tournament has a greater fitness than the winner of a smaller tournament. In many applications of this selection method, the tournament is carried out only between two individuals, i.e., a binary tournament. An important characteristic of this selection method is that there is no requirement for fitness scaling; thus, negative fitness is also allowed.

1.5.2   Crossover

In the selection process, only copies of individuals are inserted into the new population. *Crossover*, on the contrary, generates new genetic material by exchanging genetic material between individuals of a population, thus new individuals emerge. Two individuals are chosen and crossed. The resulting offspring replace the parents in the new population. Crossover manipulation can lead to the loss of "good" genetic material. Therefore, the crossover of two individuals is carried out with the probability p_c, the *crossover probability*, which is fixed before the optimization process. In each generation, $\mu \cdot p_c/2$ pairs, chosen at random, are crossed.

The crossover operation takes place as follows. A random number between zero and one is generated. If this number is smaller than the *crossover probability*, then two individuals are randomly chosen, and their chromosome pairs are split at a crossover point. The crossover point determines how the genetic material of the new individuals will be composed. For each pair of individuals, the crossover point is randomly determined anew. The representation determines how the crossover operator is applied to the individuals. In the following, the crossover operator is described for the binary and real representations.

1.5.2.1 Crossover for binary representation

When the individuals of the population are represented by binary numbers, the crossover is accomplished as follows. First, two individuals are randomly selected. Next, they are divided at the crossover point i into two parts and are newly composed. Let the two old individuals (parents) be given by:

$$\mathbf{c}_1 = \left[c_{1,1}, c_{1,2}, ..., c_{1,i}, c_{1,i+1}, ..., c_{1,n}\right]$$

$$\mathbf{c}_2 = \left[c_{2,1}, c_{2,2}, ..., c_{2,i}, c_{2,i+1}, ..., c_{2,n}\right]$$

After the crossover operation is realized at the crossover point i, two new individuals (offspring) result:

$$\mathbf{c}_1^{new} = \left[c_{1,1}, c_{1,2}, ..., c_{1,i}, c_{2,i+1}, ..., c_{2,n}\right]$$

$$\mathbf{c}_2^{new} = \left[c_{2,1}, c_{2,2}, ..., c_{2,i}, c_{1,i+1}, ..., c_{1,n}\right]$$

The new individual c_1^{new} consists of the first part of the old individual c_1 and the second part of the old individual c_2. Similarly, the second new individual c_2^{new} consists of the first part of the old individual c_2 and the second part of the old individual c_1. Figure 1.4 depicts an example.

The description above shows a one-point crossover. The extension for two-point or multipoint crossover was studied by Syswerda (1989). Also, the use of multiple individuals was investigated by Syswerda (1993). In regards to the performance of the two-point or multipoint crossover, or even the use of several individuals, depends on the problem on hand.

1.5.2.2 Crossover for real representation

In real representation, an individual consists of a real number. The use of arithmetical crossover techniques has yielded good results in the solution of constrained nonlinear optimization problems (Michalewicz, 1996). Let c_1 and

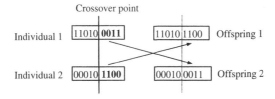

Figure 1.4 Crossover for binary representation.

c_2 be two individuals, who are to reproduce. The two offspring c_1^{new} and c_2^{new} are produced as a linear combination of their parents c_1 and c_2:

$$c_1^{new} = \lambda\, c_1 + (1-\lambda)c_2 \qquad (1.11)$$

$$c_2^{new} = (1-\lambda)\, c_1 + \lambda\, c_2 \qquad (1.12)$$

where $\lambda \in [0,1]$ is a crossover parameter.

1.5.3   Mutation

The mutation process effects a random variation upon the gene of an individual. A mutation is executed with the probability p_m, the *mutation probability*, which is fixed before the optimization. For each individual, a random number between 0 and 1 is calculated, which is compared with the mutation probability. If the random number is smaller than the probability of mutation, a gene is mutated.

1.5.3.1   Mutation for binary representation

In binary representation, one bit of a gene is selected randomly at bit position and inverted, i.e., a bit with the value 0 is turned around to a bit with the value 1, and vice versa. Figure 1.5 depicts an example for *binary mutation*.

1.5.3.2   Mutation for real representation

The mutation operator was originally developed for binary representation. Since then, other methods have been developed that allow for gene modification in real representation. These methods apply a probability distribution, which is defined over the domain of the possible values for each gene. A new value for a gene is calculated after this probability distribution. The mutation operator randomly alters one or more of the genes of a selected individual.

Let the individual $c_i = [c_{i1}, \ldots, c_{ij}, \ldots, c_{in}]$ and c_{ij} the gene to be selected for mutation. The domain of the variable c_{ij} is given by $c_{ij} = [c_{ij,min}\ c_{ij,max}]$, where $c_{ij,min}$ and $c_{ij,max}$ denote the lower and upper bounds of the variable c_{ij}, respectively. Real number mutations are of two types: uniform mutation and non-uniform mutation (Michalewicz, 1996):

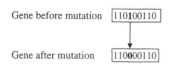

Figure 1.5 Mutation for binary representation.

1. Uniform mutation: The application of this operator results in an individual $c_i^{new} = \left[c_{i1}, ..., \tilde{c}_{ij}, ..., c_{in} \right]$ where $\tilde{c}_{ij}$ is a random value (uniform probability distribution) within the domain of c_{ij}. The mutation operator is applied with a probability p_m.

2. Nonuniform mutation: The application of this operator results in an individual $c_i^{new} = \left[c_{i1}, ..., \tilde{c}_{ij}, ..., c_{in} \right]$ where $\tilde{c}_{ij}$ is a value calculated by:

$$\tilde{c}_{ij} = \begin{cases} c_{ij} + \ddot{A}(g, c_{ijmax} - c_{ij}), & \text{if } h = 0 \\ c_{ij} + \ddot{A}(g, c_{ij} - c_{ijmin}), & \text{if } h = 1 \end{cases} \tag{1.13}$$

where h is one binary digit chosen randomly (0 or 1). The function $\Delta(g,y)$ returns a value in the interval $[0,y]$ such that the probability of $\Delta(g,y)$ initiates in zero and is increased in accordance with the generation number g such that:

$$\Delta(g,y) = y \cdot r_a \cdot \left[1 - \frac{g}{g_{max}} \right]^b \tag{1.14}$$

where r_a is a number randomly generated in the interval $[0, 1]$, b is a parameter chosen by the designer, which determines the degree of dependence on the generation number. Therefore, the search is effected uniform initially when g is small and more local in the later generations.

1.6   Genetic algorithms for optimization

1.6.1   Genetic algorithms at work

Figure 1.6 shows a GA to solve an unconstrained optimization problem. Initially, the fitness function $F(x)$ is defined, based on the problem to be solved. The probabilities p_c, p_m, and the population size μ are chosen. The individuals of the initial population P_0 are randomly initialized. So begins the first generation through the fitness calculation $F(c_i)$ with $i = 1, ..., \mu$ for each individual of the population. By applying selection to the individuals of the population P_g, a transition population $\tilde{P}$ would result. From the application of crossover with the probability p_c, a further transition from $\tilde{P}$ to population $\tilde{\tilde{P}}$ results. From the application of mutation operator, with the probability p_m, to the individuals of the population $\tilde{\tilde{P}}$ a new population results, which is designated P_{g+1}. If the maximum generation number g_{max} is not achieved, then the fitness is calculated, and the genetic operators are applied. If $g = g_{max}$, then the optimization is terminated, and the fittest individual represents the solution of the optimization problem.

By repeated application of the genetic operators, it is possible that the fittest individual of a generation was not selected or destroyed by crossover

Algorithm 3: genetic algorithms

Input: $F(x), p_c, p_m$ and μ

Output: $\mathbf{c}_i$

Auxiliary variable: g and g_{max}

begin

    $g = 0$

    initialize: $P_g = \left\{c_1, \ldots, c_i, \ldots c_\mu\right\}$

    while $\left(g \leq g_{max}\right)$ **do**

    fitness calculate: $F\left(c_i\right)$

    selection: $P_g \Rightarrow \tilde{P}$

    crossover: $\tilde{P} \Rightarrow \tilde{\tilde{P}}$

    mutation: $\tilde{\tilde{P}} \Rightarrow P_{g+1}$

    end while

  return c_i

end

Figure 1.6 GA for optimization.

or mutation. Thus, the best individual would be no more contained in the next population. This problem can be avoided by ensuring that the best individual of the previous population goes into the next generation, if the best individual of the current population has a lower fitness. The best individual is replaced only by a still better individual.

1.6.2   An optimization example

In order to check the suitability of GA, a well-known analytic function is used here. The classical GA with individuals, which are represented by bit strings, with proportionate selection, one-point crossover, and mutation by bit inversion is not always adequate and efficient to solve optimization problems (Mitchell, 1996). Here, we use a GA with real representation and the following genetic operators:

- Tournament selection with tournament size $z = 2$
- Arithmetical crossover with crossover parameter $\lambda = 0.5$
- Mutation according to a uniform probability distribution

We use the *Goldstein–Price function*. It is described by the following equation (De Jong, 1975):

$$f_G(\mathbf{x}) = \left[1 + (x_1 + x_2 + 1)^2 \, (19 - 14x_1 + 3x_1^2 - 14x_2 + 6x_1x_2 + 3x_2^2)\right] \cdot$$
$$\cdot \left[30 + (2x_1 - 3x_2)^2 \, (18 - 32x_1 + 12x_1^2 + 48x_2 - 36x_1x_2 + 27x_2^2)\right]$$

(1.15)

where $x = [x_1, x_2]^T$.

This function has only one global minimum at $x_1^* = [0, -1]^T$, with the function value $f_G(x^*) = 3$. The minimization of the *Goldstein–Price function* in Equation (1.15) is carried out by using the GA. An individual c consists of the vector $x = [x_1, x_2]^T$.

Based on Equation (1.10), the fitness for each individual is given by the following:

$$F(\mathbf{x}) = -f_G(\mathbf{x})$$

(1.16)

The GA parameters are population size $\mu = 100$, crossover probability $p_c = 0.3$, mutation probability $p_m = 0.05$, and maximum number of generations $g_{max} = 100$.

The population is randomly initialized. The minimization of the *Goldstein–Price function* during the first 20 generations is shown in Figure 1.7. The GA is able to find the global minimum of the *Goldstein–Price function* in 12 generations. On the basis of this example, it can be shown that GA are suitable for finding the minimum of highly nonlinear functions.

The next chapters of this book will present new methods using GA to automatically design optimal robust, predictive, and variable structure controllers.

1.7 Genetic programming

Genetic programming (GP) is an extension of the GA for handling complex computational structures (Howard and D'angelo, 1995; Koza, 1992). The GP

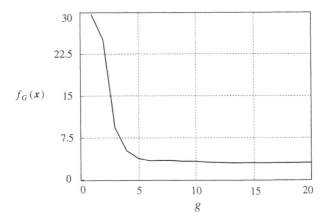

Figure 1.7 Minimization of the Goldstein–Price function using a GA.

uses a different individual representation and genetic operators and an efficient data structure for the generation of symbolic expressions, and it performs symbolic regressions. The solution of a problem by means of GP can be considered a search through combinations of symbolic expressions. Each expression is coded in a tree structure, also called a computational program, that is subdivided into nodes and presents a variable size.

The elements of GP are fixed sets of symbols, allowing the optimization of a tree structure in a more appropriate way than with only numerical parameters. The elements of the GP are divided into two alphabets: a *functional* and a *terminal*. The functional alphabet is constituted by a set of characters (+, -, *, /, sqrt, log, exp, ln, abs, and, or), that can include arithmetic operations, mathematical functions, and conditional logical operations. The terminal alphabet is a set of characters that includes constants, numbers values, and inputs appropriate for the domain of the problem. The search space is one hyperspace of all possible compositions of functions that can be recursively formed by the functional and terminal alphabets. The symbolic expressions (S-expressions) of a programming language LISP (List Processing) are a convenient way to create and to manipulate the compositions of functions and terminals. Figure 1.8 shows the expression $(x + y) \cdot \ln x$ represented as a tree structure.

The crossover operation is implemented by means of randomly selected sub-trees and performed by permutations of the sub-trees. The mutation operations usually consist of gene swapping with restrictions imposed by the designer. The optimization by means of GP can be described through a number of steps (Bäck et al., 1997):

1. Randomly create a population of trees with uniform distribution, providing symbolic expressions
2. Evaluate each tree using the fitness function
3. Apply the selection operator
4. Apply the crossover operator in a set of parent trees, chosen randomly
5. Apply the mutation operator
6. Repeat steps (2) to (5) until a stop criterion is satisfied

During the crossover operation, two trees are chosen of a similar form to the methods treated by the GA. The crossover operation must preserve the syntax of the symbolic expressions, that is, the application of the genetic

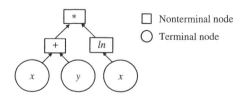

Figure 1.8 Representation in a tree structure.

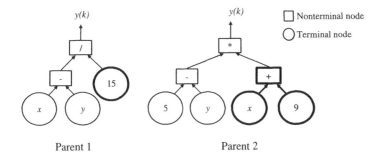

Figure 1.9 Representation of the parent trees before crossover.

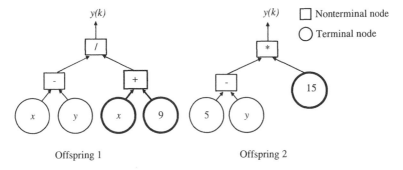

Figure 1.10 Representation of the parent trees after crossover.

operators must produce a program that can be evaluated. For instance, a sub-tree is randomly selected from a parent tree (parent no. 1) and is swapped with a randomly selected sub-tree of the parent tree (parent no. 2). The trees are then inserted into the mating pool in order to generate the offspring for the next generation. The trees to perform in the crossover operation are indicated by darker lines in Figure 1.9.

After the crossover operation is applied to the parents, the result is the creation of two offspring, as illustrated in Figure 1.10.

Mutation generates a modified tree that is copied in the next generation of the population. The mutation consists of a random modification of a function, input, or a constant in a symbolic expression of the population (Bäck et al., 1997). For example, consider the application of the mutation operator on offspring 2 shown in Figure 1.10. The terminal node with value equal to 15 is changed to the value 21, as shown in Figure 1.11.

In Chapter 8, we present a fuzzy-GP approach to mobile robot navigation and control.

1.8 Conclusions

In this chapter, an overview of genetic algorithms (GA) and genetic programming (GP) was given. The major definitions and terminology of genetic

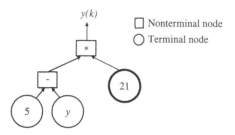

Figure 1.11 Representation of an offspring tree after mutation.

algorithms were introduced. The traditional binary and real representations, the fitness function, and the three genetic operators: selection, crossover, and mutation, were described. In order to illustrate the concepts, we showed how GA could be used to solve optimization problems. The concept of GA was extended to GP, and the representation in form of trees with new genetic operators was explained.

References

Bäck, T., *Evolutionary Algorithms in Theory and Practice*, Oxford University Press, Oxford, 1996.

Bäck , T., Fogel, D.B., and Michalewicz, Z., *Handbook of Evolutionary Computation*, Institute of Physics Publishing, Philadelphia and Oxford University Press, New York, Oxford, 1997.

Bethke, A.D., *Genetic Algorithms as Function Optimization*, Ph.D. dissertation, University of Michigan, 1981.

Blickle, T., *Theory of Evolutionary Algorithms and Applications to System Synthesis*, Ph.D. dissertation, Eidgenössische Technische Hochschule, Zürich, 1997.

Davis, L. (Ed.), *Handbook of Genetic Algorithms*, Van Nostrand Reinhold, New York, 1991.

De Jong, K.A., *An Analysis of the Behavior of a Class of Genetic Adaptive Systems*, Ph.D. dissertation, University of Michigan, 1975.

Fogel, D.B., *Evolutionary Computation: Toward a New Philosophy of Machine Intelligence*, IEEE Press, Piscataway, 1995.

Goldberg, D.E., *Genetic Algorithms in Search, Optimization and Machine Learning*, Addison Wesley, Reading, 1989.

Goldberg, D.E., Deb, K., and Thierens, D., Toward a better understanding of mixing in genetic algorithms, *Journal of the Society of Instrument and Control Engineers*, 32, 1, 10–16, 1993.

Holland, J.H., *Adaptation in Natural and Artificial Systems*, University of Michigan Press, Ann Arbor, 1975.

Howard, L.M. and D'angelo, D.J., The GA-P: A genetic algorithm and genetic programming hybrid, *IEEE Expert*, 10, 3, 11–15, 1995.

Koza, J.R., *Genetic Programming: On the Programming of Computers by Means of Natural Selection*, MIT Press, Cambridge, 1992.

Michalewicz, Z., *Genetic Algorithms + Data Structure = Evolution Programs*, Springer-Verlag, Berlin, 1996.

Miller, B.L. and Goldberg, D.E., *Genetic Algorithms, Tournament Selection and the Effects of Noise*, IlliGAL Report No. 95006, Department of General Engineering, University of Illinois, 1995.

Mitchell, M., *An Introduction to Genetic Algorithms*, MIT Press, Cambridge, 1996.

Syswerda, G., Uniform crossover in genetic algorithms, in *Proceedings of the 3rd International Conference on Genetic Algorithms*, Morgan Kaufmann Publishers, Los Altos, 1989, pp. 2–9.

Syswerda, G., Simulated crossover in genetic algorithms, in *Foundations of Genetic Algorithms*, Whitley, L.D., Ed., Morgan Kaufmann Publishers, Los Altos, 1993, pp. 239–255.

Wright, A.H., Genetic algorithms for real parameter optimization, in *Foundations of Genetic Algorithms*, Rawlins, G., Ed., Morgan Kaufmann Publishers, Los Altos, pp. 205-218, 1991.

chapter two

Optimal robust control

2.1 Introduction to the control theory

Control theory aims to develop methods for analysis and design of control systems. A simple control system consists of the controller and the plant. Electrical motors, machine tools, and airplanes are examples of plants. The controller obtains information about the plant by means of sensors. The information is processed, and a control signal $u(t)$ is calculated, in order to influence the dynamic behavior of the plant. The goal of the control, despite disturbance $\delta(t)$ acting on the plant, is to keep the value of the controlled variable (the output variable) $y(t)$ within tolerance of the value given by the reference variable (set-point) $r(t)$. Figure 2.1 shows an elementary control system.

A fundamental question is how to design the controller. Generally, a model of the plant is necessary in order to design a controller. By using the classical methods of control theory, the behavior of the plant is described by a mathematical model. It can be obtained by application of physical laws or by means of experimental identification.

Finding the mathematical model applying physical laws yields mostly a system of nonlinear differential equations. A system of linear time-invariant differential equations with constant coefficients can be obtained by linearization around a point. Applying the Laplace Transformation to the equation systems with initial values set to zero yields a plant model in the form of a transfer function. This way, the methods from linear systems after Nyquist or Bode can be employed.

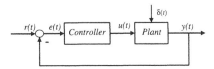

Figure 2.1 Control system.

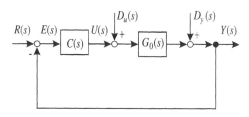

Figure 2.2 Block diagram of the control system.

If input–output data from the plant is available, a mathematical model can be obtained by means of experimental identification. Starting from a linear time-invariant model with a known structure, the parameters of the model can be determined using the measured data of the plant.

We now consider a linear time-invariant single-input–single-output (SISO) system. A continuous time-invariant linear model describes the plant. The control system as shown in Figure 2.2 has one output variable (the controlled variable) $Y(s)$, the reference variable (set point) $R(s)$, the input disturbance variable $D_u(s)$, and the output disturbance variable $D_y(s)$; where

$R(s)$	Reference variable
$D_u(s)$	Disturbance at the plant input
$D_y(s)$	Disturbance at the plant output
$Y(s)$	Output variable
$U(s)$	Control signal
$E(s)$	Error
$C(s)$	Transfer function of the controller
$G_0(s)$	Nominal transfer function of the plant

The error $E(s)$ for the control system shown in Figure 2.2 is given by

$$E(s) = \frac{1}{1+C(s)G_0(s)} \cdot (R(s) - D_y(s)) - \frac{G_0(s)}{1+C(s)G_0(s)} \cdot D_u(s) \qquad (2.1)$$

The open loop transfer function $L(s)$ is defined as follows:

$$L(s) = C(s)G_0(s) \qquad (2.2)$$

The sensitivity function is defined as:

$$S(s) = \left. \frac{E(s)}{R(s)} \right|_{\substack{D_u(s)=0 \\ D_y(s)=0}} = \frac{1}{1+L(s)} \qquad (2.3)$$

The complementary sensitivity function is defined as follows:

$$T(s) = \left. \frac{Y(s)}{R(s)} \right|_{\substack{D_u(s)=0 \\ D_y(s)=0}} = \frac{C(s)G_0(s)}{1+C(s)G_0(s)} \qquad (2.4)$$

The first important issue in designing a control system is the consideration of stability. The stability of the closed loop can be determined through the roots of the characteristic equation:

$$1+C(s)G_0(s) = 0 \qquad (2.5)$$

A control system is stable if and only if all roots of the characteristic equation are situated in the left half of the s-plane. If its real parts are negative, it displays absolute stability. According to the Hurwitz test, the absolute stability of a control system can be tested by means of the coefficients of the characteristic equation, without calculation of the exact position of the roots of the characteristic equation.

A simplified model of the plant is generally used in design or analysis of a control system. The model usually contains errors. The causes for such errors are as follows:

- Deviation between the real parameters and the modeled parameters
- Modifications of the plant parameters by means of age, environment conditions, and dependencies on the work point
- Errors by simplification of the model

The model error stands for model uncertainty. It is of fundamental importance that in designing a controller, the model uncertainty be considered, so that the stability of the control system can be guaranteed. In this case, it is called robust stability of the control system. If, during the design, the model uncertainty is considered, then it is called the design of the robust controller (Mueller, 1996).

When designing robust controllers, the model uncertainty of the plant is explicitly considered. There are two kinds of model uncertainty: *structured and nonstructured model uncertainty*. Structured model uncertainty or parametric model uncertainty is caused by parameter modifications of the plant, and it can be described using interval methods (Ackermann, 1993; Bhattacharyya et al., 1995). The cause of nonstructured model uncertainty is usually nonlinearities of the plant or modifications of the work point. This type of model uncertainty can be described using the H_∞- theory (Djaferis, 1995; Doyle et al., 1992).

Classical methods for controller design use a nominal model of the plant. The robustness of the control loop is indicated by the parameters: phase margin and gain margin (Djaferis, 1995). Methods for the design of robust controllers, based on the H_∞- theory, use a family of models of the plant. In this case, a nominal model and the model uncertainty are considered. It is necessary to guarantee the stability of the control loop by taking into account the model uncertainty. To describe the conditions for robust stability and disturbance rejection of the control system, the H_∞- norm will be used.

2.2   Norms of signals and functions

In this subsection, norms of signals and functions are described. A norm evaluates elements of a metric space by a real, positive number that represents a measure for the size of the element (Mueller, 1996). The norm refers to vector-valued signals, to real functions of time, or to the function of the variable s of the Laplace Transformation. Commonly used norms in automatic control are the Euclidean norm and the Maximum norm (Tschebyschew norm) (Boyd and Barrat, 1991).

The L_2-norm* of the signal $v(t)$ is defined by (Doyle et al., 1992):

$$\|v\|_2 = \left(\int_{-\infty}^{\infty} (v(t))^2 \, dt \right)^{0.5}$$

(2.6)

and the L_∞- norm is defined by the following (Doyle et al., 1992):

$$\|v\|_\infty = \max_t |v(t)|$$

(2.7)

The L_∞- norm of $v(t)$ is the maximum amplitude of the signal $v(t)$.

The H_2- norm** of the transfer function $G(s)$*** is defined by the following (Doyle et al., 1992):

$$\|G\|_2 = \left(\frac{1}{2\pi} \int_{-\infty}^{\infty} |G(jw)|^2 \, dw \right)^{0.5}$$

(2.8)

and the H_∞- norm is defined by the following (Doyle et al., 1992):

$$\|G\|_\infty = \max_w |G(jw)|$$

(2.9)

The H_∞- norm of $G(s)$ represents the maximum amplitude on the Bode magnitude plot. Figure 2.3 shows the magnitude plot for the following transfer function:

---

* L is an abbreviation for the mathematician Lebesgue (Boyd and Barrat, 1991).
** H is an abbreviation for the mathematician Hardy (Boyd and Barrat, 1991).
*** j stands for the imaginary number, i.e., j= $\sqrt{-1}$

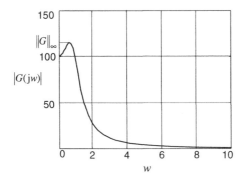

Figure 2.3 Bode magnitude plot of $G(s)$.

$$G(s) = \frac{100}{s^2 + s + 1} \tag{2.10}$$

Norms are usually used in connection to optimization problems, i.e., minimization of the sensitivity function, e.g., min of $\|S\|_\infty$ (Zames, 1981). For the optimization problem, with a norm that will be optimized, the transfer function is required to remain stable. In the following will be given two definitions. Assume that the rational transfer function $G(s)$ is stable:

   Definition 2.1 (Mueller, 1996): The transfer function $G(s)$ is characterized as proper, if $G(j\infty) < \infty$.
   Definition 2.2 (Mueller, 1996): The transfer function $G(s)$ is characterized as strictly proper, if $G(j\infty) = 0$.

A relationship between the L_2- norm and the H_∞- norm is given by the following theorem:
   Theorem 2.1 (Vidyasagar, 1985): If $\|v\|_2 \leq \infty$, and $Y(s) = G(s)V(s)$, where $G(s)$ is a stable transfer function without poles on the imaginary axis, then

$$\max_{v(t) \in L_2} \frac{\|y\|_2}{\|v\|_2} = \|G\|_\infty \tag{2.11}$$

A strong mathematical treatment of norms can be found in Vidyasagar (1985).

2.3 *Description of model uncertainty*

A condition for robust stability of a control system can be described, taking into account plant uncertainty. Methods for identification of the plant model, taking into account the uncertainties by means of the L_∞- norm or the H_∞- norm have been examined. Such methods are known as robust identification methods (Milanese et al., 1996). These methods consist of the identification of a mathematical model, which fits the experimental data of the plant to the L_∞- norm or to the H_∞- norm. This subsection presents explicit models to describe the nonstructured uncertainty. The two most frequent models used for describing plant behavior are the multiplicative and the additive.

By using the multiplicative model, the transfer function of the real (perturbed) plant $G(s)$ is described by the following (Doyle et al., 1992):

$$G(s) = G_0(s)\big(1 + \Delta(s)W_m(s)\big) \tag{2.12}$$

where
$G_0(s)$ Nominal transfer function
$\Delta(s)$ Disturbance (perturbation) acting on the plant
$W_m(s)$ Weighting function that represents an upper bound of the multiplicative uncertainty

By using the additive model, the transfer function of the real plant $G(s)$ is described by the following (Doyle et al., 1992):

$$G(s) = G_0(s) + \Delta(s)W_a(s) \tag{2.13}$$

where $W_a(s)$ is the weighting function that represents an upper bound of the additive uncertainty.

Determination of uncertainty can be realized by means of experimental identification. Let the plant be described by Equation (2.12). This equation can be transformed to:

$$\frac{G(s)}{G_0(s)} - 1 = \Delta(s)W_m(s) \tag{2.14}$$

Suppose that the perturbation that actuates upon the plant is unknown but bounded, i.e., $||\Delta(s)||_\infty \leq 1$, then results with $s = jw$ are as follows:

$$\left|\frac{G(jw)}{G_0(jw)} - 1\right| \leq \left|W_m(jw)\right| \qquad \forall w \tag{2.15}$$

Thus, the model uncertainty is represented by the deviation between the normalized plant and one (unity).

The objective of the methods for robust identification is to determine an upper bound of the model uncertainty $W_m(s)$. This can be carried out as follows (Djaferis, 1995; Doyle et al., 1992): It is assumed that the plant is stable, and its transfer function is determined by means of experiments in the frequency domain. The plant is excited with a sinusoidal input signal at different frequencies w_i with $i = 1,...,m$. For each frequency, the respective amplitude and phase are measured. The experiment is repeated n times.

Let the amplitude $M_r(w_i)$ and the phase $\varphi_r(w_i)$ of the r-th experiment in the frequency w_i. By using the pairs (M_i, φ_i) for each frequency w_i, the nominal transfer function of the plant $G_0(s)$ is determined, where $M_i = |G_0(jw_i)|$, and $\varphi_i = \arg G_0(jw_i)$. Next, the weighting function $W_m(jw_i)$ is selected according to Equation (2.15). For each frequency w_i, the values:

$$\left| \frac{M_r(w_i)\, e^{j\varphi_r(w_i)}}{M_i\, e^{j\varphi_i}} - 1 \right| \quad \text{with } i = 1,...,m \text{ und } r = 1,...,n$$

represent the deviation between the normalized plant and one. Thus, an upper bound of the model uncertainty can be determined (Doyle et al., 1992).

A method for the determination of an upper bound for additive and multiplicative uncertainty is presented in Tan and Li (1996). If the input–output values are available in the time domain, then these data can be transformed into the frequency domain using the Fourier Transformation. Methods for robust identification are discussed in Milanese et al. (1996) and Smith and Dahleh (1994).

2.4   Robust stability and disturbance rejection

This section describes a condition for robust stability of a control system, taking into account the plant uncertainty, and a condition for disturbance rejection of a control system, taking into account the disturbance that acts on the plant.

2.4.1   Condition for robust stability

Consider the control system shown in Figure 2.4. The controller is described by means of a transfer function with fixed structure $C(s,k)$. The vector k assigns the vector of the controller parameters:

$$k = \left[k_1, k_2, ..., k_m \right]^{\mathrm{T}} \tag{2.16}$$

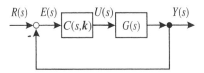

Figure 2.4 Control system composed of a controller with fixed structure, and a plant with model uncertainty.

The plant is described by a multiplicative model according to Equation (2.12). It is assumed that the model uncertainty $W_m(s)$ is stable and bounded, and that no unstable poles of $G_0(s)$ are canceled in forming $G(s)$.

The condition for robust stability is stated as follows (Doyle et al., 1992): If the nominal control system ($\Delta(s) = 0$) is stable with the controller $C(s,k)$, then the controller $C(s,k)$ guarantees robust stability of the control system, if and only if the following condition is satisfied:

$$\left\| \frac{C(s,k)G_0(s)W_m(s)}{1+C(s,k)G_0(s)} \right\|_\infty < 1 \qquad (2.17)$$

This condition for robust stability represents only a sufficient condition. So, the robust stability of a control system can be evaluated by means of the H_∞-norm.

If the plant is described by an additive model according to Equation (2.13), and it is assumed that the uncertainty $W_a(s)$ is stable and bounded, then the condition for robust stability is stated as follows (Doyle et al., 1992): If the nominal control system ($\Delta(s) = 0$) is stable with the controller $C(s,k)$, then the controller $C(s,k)$ guarantees robust stability of the control system, if and only if the following condition is satisfied:

$$\left\| \frac{C(s,k)W_a(s)}{1+C(s,k)G_0(s)} \right\|_\infty < 1 \qquad (2.18)$$

Generally, the multiplicative model is used. If the plant is described by an additive model, it can be easily converted into a multiplicative model. This text will consider the multiplicative model.

Applying the definition of the H_∞- norm, according to Equation (2.9), on the condition for robust stability, results in the following:

$$\left\| \frac{C(s,k)G_0(s)W_m(s)}{1+C(s,k)G_0(s)} \right\|_\infty =$$

$$= \max_{w \in [0,\infty)} \left(\frac{\big(C(jw,k)G_0(jw)W_m(jw)\big)\big(C(-jw,k)G_0(-jw)W_m(-jw)\big)}{\big(1+C(jw,k)G_0(jw)\big)\big(1+C(-jw,k)G_0(-jw)\big)} \right)^{0.5}$$

$$= \max_{w \in [0,\infty)} (\alpha(w,k))^{0.5} \qquad (2.19)$$

where

$$\alpha(w,k) = \frac{\alpha_z(w,k)}{\alpha_n(w,k)} = \frac{\big(C(jw,k)G_0(jw)W_m(jw)\big)\big(C(-jw,k)G_0(-jw)\,W_m(-jw)\big)}{\big(1+C(jw,k)G_0(jw)\big)\big(1+C(-jw,k)G_0(-jw)\big)} \qquad (2.20)$$

Thus, the condition for robust stability in the frequency domain is represented by the following:

$$\max_{w \in [0,\infty)} (\alpha\,(w,k))^{0.5} < 1 \qquad (2.21)$$

The function $\alpha(w,k)$ in Equation (2.20) can also be expressed in the following form:

$$\alpha(w,k) = \frac{\alpha_z(w,k)}{\alpha_n(w,k)} = \frac{\displaystyle\sum_{j=0}^{p} \alpha_{zj}(k)w^{2j}}{\displaystyle\sum_{i=0}^{q} \alpha_{ni}(k)w^{2i}} \qquad (2.22)$$

Both polynomials $\alpha_z(w,k)$ and $\alpha_n(w,k)$ have only even powers of w, and the coefficients $\alpha_{zj}(k)$ and $\alpha_{ni}(k)$ are functions of k. Considering the robust stability condition, a requirement for the existence of a controller concerning its structure can be described as follows: In Equation (2.22), the degree p of $\alpha_z(w,k)$ must be smaller than the degree q of $\alpha_n(w,k)$, i.e., $p < q$, so that the function $\alpha(w,k)$ has a finite value for $w \geq 0$ and tends toward zero for $w \to \infty$. This implies that the product of $C(s,k) \times G_0(s)$ and $W_m(s)$ must be strictly proper and proper rational functions, respectively.

2.4.2 Condition for disturbance rejection

Disturbances with deterministic signal form, e.g., step function, sinusoidal function, are assumed in the classical methods for controller design (Åström and Hägglund, 1994). Taking into account disturbances using the H_∞- norm, the type of the signal can be arbitrary. It must be assumed, however, that the amplitude of the signal is bounded. In the following, the condition for disturbance rejection will be described.

Consider the control system shown in Figure 2.5 with disturbance $D_y(s)$ acting at the plant output. The controller with fixed structure is given by a rational transfer function $C(s,k)$. The plant is described by its nominal transfer function $G_0(s)$.

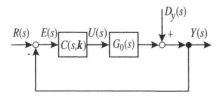

Figure 2.5 Control system with disturbance acting on the plant output.

Let the reference signal $R(s) = 0$, then the relation of the controlled variable $Y(s)$ to the disturbance at the output $D_y(s)$ can be described as follows:

$$\frac{Y(s)}{D_y(s)} = \frac{1}{1 + C(s,k)G_0(s)} \tag{2.23}$$

Applying Theorem 2.1 to Equation (2.23) yields:

$$\max_{d_y(t) \in L_2} \frac{\|y\|_2}{\|d_y\|_2} = \left\| \frac{1}{1 + C(s,k)G_0(s)} \right\|_\infty \tag{2.24}$$

The condition for disturbance rejection indicates that the maximal amplitude of the output variable $y(t)$ caused by means of the disturbance on the plant output $d_y(t)$, should not exceed a pre-fixed upper bound γ, i.e.,

$$\max_{d_y(t) \in L_2} \frac{\|y\|_2}{\|d_y\|_2} = \left\| \frac{1}{1 + C(s,k)G_0(s)} \right\|_\infty < \gamma \tag{2.25}$$

where $\gamma \leq 1$ is a design parameter.

According to Chen et al. (1995), the introduction of a weighting function $W_d(s)$ that consists of low-pass filter into Equation (2.25) yields:

$$\left\| \frac{W_d(s)}{1 + C(s,k)G_0(s)} \right\|_\infty < \gamma \tag{2.26}$$

This condition for disturbance rejection represents only a sufficient condition. The disturbance rejection of a control system can be evaluated by using H_∞- norm.

Applying the definition of the H_∞- norm according to Equation (2.9) on the condition for disturbance rejection results in the following:

$$\left\| \frac{W_d(s)}{1 + C(s,k)G_0(s)} \right\|_\infty =$$

$$= \max_{w \in [0,\infty)} \left(\frac{W_d(jw)W_d(-jw)}{\left(1 + C(jw,k)G_0(jw)\right)\left(1 + C(-jw,k)G_0(-jw)\right)} \right)^{0.5}$$

$$= \max_{w \in [0,\infty)} (\beta(w,k))^{0.5} \tag{2.27}$$

where

$$\beta(w,k) = \frac{\beta_z(w,k)}{\beta_n(w,k)} = \frac{W_d(jw)W_d(-jw)}{\left(1 + C(jw,k)G_0(jw)\right)\left(1 + C(-jw,k)G_0(-jw)\right)} \quad (2.28)$$

So, the condition for disturbance rejection in the frequency domain is represented by:

$$\max_{w \in [0,\infty)} (\beta(w,k))^{0.5} < \gamma \quad (2.29)$$

The function $\beta(w,k)$ in Equation (2.28) can also be expressed in the following form:

$$\beta(w,k) = \frac{\beta_z(w,k)}{\beta_n(w,k)} = \frac{\sum\limits_{j=0}^{r} \beta_{zj}(k)w^{2j}}{\sum\limits_{i=0}^{q} \beta_{ni}(k)w^{2i}} \quad (2.30)$$

Both polynomials $\beta_z(w,k)$ and $\beta_n(w,k)$ have only even powers of w, and the coefficients $\beta_{zj}(k)$ and $\beta_{ni}(k)$ are functions of k. Considering the disturbance rejection condition, a requirement for the existence of a controller concerning its structure can be described as follows: In Equation (2.30), the degree p of $\beta_z(w,k)$ must be smaller than the degree q of $\beta_n(w,k)$, i.e., $p < q$, so that the function $\beta(w,k)$ has a finite value for $w \geq 0$ and tends toward zero for $w \to \infty$. This implies that the product of $C(s,k) \times G_0(s)$ and $W_d(s)$ must be strictly proper and proper rational functions, respectively.

2.5 Controller design

Robust controller design can be carried out by well-known design methods using the H_∞- theory (Khargonekar and Rotea, 1991; Kwaakernaak, 1993; Doyle et al., 1994; Bernstein and Haddad, 1994; Snaizer, 1994). One of the largest disadvantages of these design methods is that the structure of the controller cannot be fixed a priori. Moreover, these design methods provide controllers with high order, which are difficult to implement. Therefore, it is necessary to apply order reduction methods to obtain controllers with reduced order. In this section, the design problem is conditioned in such a way that the structure of the controller can be given a priori. Thus, PID automatic controllers can then be designed using the H_∞- theory.

The last section described a condition for robust stability that ensures stable operation of the control system subject to plant perturbation, but no statement was given about the tracking behavior of the control systems. In many cases, however, a good tracking behavior is desired. The design of controllers generally occurs in two steps: first, the selection of an appropriate controller structure and second, the determination of suitable controller parameters. The selection of an appropriate controller structure depends on the plant and technical implementation details. The determination of suitable controller parameters depends on the requirements of the control system. Typical requirements are: short settling time, small overshoot, good damping or small value of the squared error surface. Here, the determination of the controller parameters is carried out by the minimization of the performance index integral of the squared error (ISE) and the integral of the time-weighted squared error (ITSE) (Westcott, 1954; Schneider, 1966).

Because the condition for robust stability and the condition for disturbance rejection are described in the frequency domain, the controller design is carried out in the frequency domain. So, the performance index must be computable in the frequency domain. In this section, optimal controllers are thus designed by the minimization of the performance index squared error and the time-weighted squared error.

2.5.1 Optimal controller design

Consider the control system shown in Figure 2.6. The controller is described by the transfer function $C(s,k)$, and the plant is described by the nominal transfer function $G_0(s)$. It is assumed that good tracking behavior of the control system is essential and, therefore, will be optimized.

The performance index, the ISE J, is given by:

$$J = \int_0^\infty (e(t))^2 \, dt \tag{2.31}$$

It can be described in the frequency domain by means of the Parseval Theorem (Jury, 1974):

$$J = -\frac{1}{2\pi j} \int_{-j\infty}^{+j\infty} E(s)E(-s) \, ds \tag{2.32}$$

The error $E(s)$ for the control system shown in Figure 2.6 is given by:

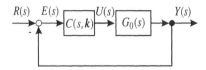

Figure 2.6 Control system.

$$E(s) = \frac{1}{1 + C(s,k)G_0(s)} \cdot R(s) \qquad (2.33)$$

The reference signal (set point) is an unit step function given by:

$$R(s) = \frac{1}{s} \qquad (2.34)$$

The error $E(s)$ can be expressed then as a rational function:

$$E(s) = \frac{D(s)}{A(s)} = \frac{\sum\limits_{j=0}^{m} d_j s^{m-j}}{\sum\limits_{i=0}^{n} a_i s^{n-i}} \qquad (2.35)$$

In this case, the degree m of the polynomial $D(s)$ must be smaller than the degree n of the polynomial $A(s)$, so that the squared error J in Equation (2.32) has a finite value. Moreover, a_n can also be zero.

Introducing the error $E(s)$ from Equation (2.35) into Equation (2.32) results in the following:

$$J_n = -\frac{1}{2\pi j} \int\limits_{-j\infty}^{+j\infty} \frac{\left(\sum\limits_{j=0}^{m} d_j s^{m-j}\right)\left(\sum\limits_{j=0}^{m} d_j (-s)^{m-j}\right)}{\left(\sum\limits_{i=0}^{n} a_i s^{n-i}\right)\left(\sum\limits_{i=0}^{n} a_i (-s)^{n-i}\right)} ds \qquad (2.36)$$

Equation (2.36) can be solved analytically by means of the Residue Theorem.

A disadvantage of this performance index is that its minimization can result in a tracking behavior with small overshoot but a long settling time. An improvement of the tracking behavior can be obtained by using the integral of the time-weighted squared error (ITSE) (Westcott, 1954), which is given by:

$$I = \int\limits_{0}^{\infty} t \left(e(t)\right)^2 dt \qquad (2.37)$$

Using the Parseval Theorem (Jury, 1974):

$$\int\limits_{0}^{\infty} f(t)h(t)dt = \frac{1}{2\pi j} \int\limits_{-j\infty}^{+j\infty} F(s)H(-s)ds \qquad (2.38)$$

It can be shown that the integral in Equation (2.37) can be solved analytically in the frequency domain. Set $h(t) = e(t)$ and $f(t) = te(t)$. Applying the Laplace Transformation yields $H(s) = E(s)$, and according to the Differentiation Theorem, $F(s) = -dE(s)/ds$. This way, the time-weighted squared error I in the frequency domain is given by:

$$I = -\frac{1}{2\pi j}\int_{-j\infty}^{+j\infty}\frac{d}{ds}(E(s))\cdot E(-s)ds \qquad (2.39)$$

Introducing the error $E(s)$ from Equation (2.35) into Equation (2.39) results in the following:

$$I_n = -\frac{1}{2\pi j}\int_{-j\infty}^{+j\infty}\frac{d}{ds}\left\{\frac{\sum_{j=0}^{m}d_j s^{m-j}}{\sum_{i=0}^{n}a_i s^{n-i}}\right\}\left(\frac{\sum_{j=0}^{m}d_j(-s)^{m-j}}{\sum_{i=0}^{n}a_i(-s)^{n-i}}\right)ds \qquad (2.40)$$

Equation (2.40) can also be solved by means of the Residue Theorem. Closed solutions for J_n, according to Equation (2.36), and for I_n, according to Equation (2.40) which depends on the coefficients a_i with $i = 0,...n$ and d_j with $j = 0,...m$, can be found in Schneider (1966) and Westcott (1954). An example for $n = 5$ is given in the next chapter by Equations (3.9) and (3.10). Analysis of these equations shows that the squared error and the time-weighted squared error are nonlinear functions of the coefficients of the numerator and denominator of Equation (2.35).

If the control system is stable, then the values of J_n and I_n are always positive. For unstable control systems, the values of J_n and I_n are not computable. Because the coefficients a_i with $i = 0,...n$ and d_j with $j = 0,...m$ contain the controller parameters, the calculation of suitable controller parameters can be carried out by minimization of $J_n(k)$ or $I_n(k)$.

Formulations of the design of optimal robust controllers and optimal disturbance rejection controllers with fixed structures are given below. These designs are based on the minimization of the performance index, ISE, and ITSE, taking into account the robust stability or the disturbance rejection.

2.5.2   Optimal robust controller design

In designing optimal robust controllers with fixed structure, both the tracking behavior and the robust stability are considered. The controller design is formulated as a constraint optimization problem, i.e.:

$$\min_{k} J_n(k) \quad \text{subject to} \quad \max_{w}\left(\alpha(w,k)\right)^{0.5} < 1$$

or

$$\min_{k} I_n(k) \quad \text{subject to} \quad \max_{w} \left(\alpha(w,k)\right)^{0.5} < 1$$

The optimization problem consists of the minimization of the performance index ISE $J_n(k)$ or the ITSE $I_n(k)$ subject to the robust stability constraint $\max(\alpha(w,k))^{0.5} < 1$. The objective of the optimization is to determine the vector of the controller parameters k^*, so that the value of the performance index $J_n(k^*)$ or $I_n(k^*)$ is minimum and the condition for robust stability $\max(\alpha(w,k^*))^{0.5} < 1$ is satisfied.

2.5.3   Optimal disturbance rejection controller design

In designing optimal disturbance rejection controllers with fixed structure, both the tracking behavior and the disturbance rejection are considered. The controller design is formulated as a constraint optimization problem, i.e.:

$$\min_{k} J_n(k) \quad \text{subject to} \quad \max_{w} \left(\beta(w,k)\right)^{0.5} < \gamma$$

or

$$\min_{k} I_n(k) \quad \text{subject to} \quad \max_{w} \left(\beta(w,k)\right)^{0.5} < \gamma$$

The optimization problem consists of the minimization of the performance index ISE $J_n(k)$ or the ITSE $I_n(k)$ subject to the disturbance rejection constraint $\max(\beta(w,k))^{0.5} < \gamma$. The objective of the optimization is to determine the vector of the controller parameters k^* so that the value of the performance index $J_n(k^*)$ or $I_n(k^*)$ is minimum and the condition for disturbance rejection $\max(\beta(w,k^*))^{0.5} < \gamma$ is satisfied.

Because the design problems by optimal robust controllers with fixed structure as well as by optimal disturbance rejection controllers with fixed structure consists of the solution of a nonlinear optimization problem with a constraint, they are described in the next section, along with some definitions concerning optimization.

2.6   Optimization

2.6.1   The optimization problem

In general, the optimization problem is defined* as follows (Dixon and Szegö, 1978; Michalewicz, 1996):

* The limitation on a minimization problem does not mean a restriction generally, because a maximization problem can be converted into a minimization problem, as follows:

$$\max_{x \in \mathfrak{R}^n} f(\mathbf{x}) = -\min_{x \in \mathfrak{R}^n} [-f(\mathbf{x})]$$

$$\min_{x \in \Re^n} f(\mathbf{x}) \quad \text{subject to} \quad \begin{cases} g_j(\mathbf{x}) \le 0 & \text{with } j = 1,...,m \\ h_i(\mathbf{x}) = 0 & \text{with } i = m+1,...,r \end{cases}$$

The function $f(x)$ to be optimized is called the objective function. To it applies:

$$f: \Re^n \to \Re$$

The objective function $f(x)$ assigns to each vector x a scalar value. The vector x consists of n variables:

$$\mathbf{x} = \begin{bmatrix} x_1, & x_2,..., x_n \end{bmatrix}^{\mathrm{T}} \in \Re^n \qquad (2.41)$$

The set $S \subseteq \Re^n$ designates the search space and is an n-dimensional parallelepiped in $\Re^n$, which is defined by the lower and upper bounds of the variables:

$$x_{i\,\min} \le x_i \le x_{i\,\max} \quad \text{with } i = 1,...,n$$

The set $Z \subseteq S$ is called the feasible set and is defined by the set S and the constraints. For optimization problems, it is assumed that a set of variables $x = [x_1, x_2,...,x_n]^{\mathrm{T}}$ can be interpreted as a point in $\Re^n$. A point $x \in S \subseteq \Re^n$ that satisfies the constraints is called a feasible point. If there are no constraints, the optimization problem is called unconstraint, therefore, $Z = S$.

A feasible point x^* is called a local point of minimum, and its function value $f(x^*)$ is a local minimum for the optimization problem if and only if there exists a $\varepsilon > 0$, such that $f(x^*) \le f(x)$ for all $x \in Z$ with $|x - x^*| < \varepsilon$, where ε is a small, real, positive number.

A feasible point x^* is called a global point of minimum, and its function value $f(x^*)$ is a global minimum for the optimization problem if and only if $f(x^*) \le f(x)$ for all $x \in Z$.

An objective function is called unimodal if it has only a local minimum. Otherwise, it is called multimodal. Global optimization methods aim to find both the optimal value of an objective function and the respective optimal values of the variables. Figure 2.7 shows an example of a one-dimensional multimodal objective function $f(x_1)$ with two local minima and a global minimum in the domain of the search space. Because controller design consists of the solution of an optimization problem with a constraint, a brief description of methods to handle constraints is given.

2.6.2 *Constraint handling*

The search space S generally consists of two subspaces: the feasible Z and the unfeasible U. Figure 2.8 shows a possible search space

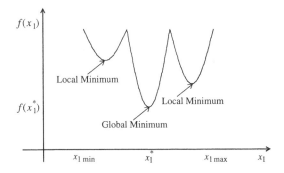

Figure 2.7 Example of a one-dimensional multimodal function.

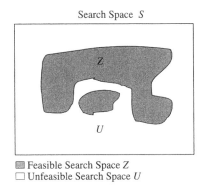

Feasible Search Space Z
Unfeasible Search Space U

Figure 2.8 Search space.

(Michalewicz, 1995). In this text, no assumption will be made about the search space.

For unconstrained optimization problems, all individuals are considered feasible, but for constraint optimization problems feasible and unfeasible individuals must be differentiated. The goal is, therefore, to find the feasible global optimum. There is no general method for handling constraints. The most frequently used methods are penalty functions (Homaifar et al., 1994; Joines and Houck, 1994; Michalewicz, 1995).

A constraint optimization problem can be transformed into an unconstrained optimization problem by means of addition of a penalty function, which considers the violation of the constraints. The optimization problem can be described as follows:

$$\min_{\mathbf{x} \in \Re^n} \ \left(f(\mathbf{x}) + P(\mathbf{x}) \right)$$

where $P(x)$ is a penalty function. If no violation of the constraints occurs, i.e., if all the constraints are satisfied, then $P(x)$ is set to zero; otherwise, $P(x)$ assumes a positive value. Usually, the penalty function is defined as follows (Kim and Myung, 1997; Michalewicz, 1995):

$$P(\mathbf{x}) = M_s \left(\sum_{j=1}^{m} (g_j^+(\mathbf{x}))^2 + \sum_{i=m+1}^{r} (h_i(\mathbf{x}))^2 \right) \tag{2.42}$$

where
M_s is a positive penalty parameter
$g_j^+(\mathbf{x}) = \max(0, g_j(\mathbf{x}))$, for $1 \le j \le m$
$h_i(\mathbf{x})$, for $m + 1 \le i \le r$

By taking in account constraints using penalty functions, unfeasible individuals are penalized by means of a fitness reduction. The fitness function depends both on the objective function $f(\mathbf{x})$ and on the penalty function $P(\mathbf{x})$. The penalty function reflects the violation of the constraints. Unfeasible individuals are punished by the penalty function, so that the optimization is forced in the direction of the feasible subspace.

The fitness function $F(\mathbf{x})$ is defined as follows:

$$F(\mathbf{x}) = \begin{cases} -f(\mathbf{x}), & \text{if } \mathbf{x} \in Z \\ -(f(\mathbf{x}) + P(\mathbf{x})), & \text{if } \mathbf{x} \in U \\ -M_t, & \text{if } f(\mathbf{x}) \text{ does not exist} \end{cases} \tag{2.43}$$

The penalty value is zero, if an individual lies in the feasible space Z, thus if an individual violates no constraint. If an individual is in the search space S, but one or more constraints are not satisfied, then a penalty function $P(\mathbf{x})$ is added to the objective function, resulting in a punished fitness. If the objective function is not computable, then the fitness is assigned the penalty parameter M_t.

2.7   Conclusions

In this chapter, the design of optimal robust controllers and optimal disturbance rejection controllers, both with fixed structure, were formulated. The definitions of norms of signals and functions were introduced. The two most common models to describe plant uncertainty, additive and multiplicative, were presented. By means of the H_∞- norm, two conditions were described: one for robust stability and one for disturbance rejection. The design of optimal robust controllers and the design of optimal disturbance controllers, both with fixed structure, were formulated as a constrained optimization problem. The problem consists of the minimization of a performance index (the integral of the squared error or the integral of the time-weighted squared error) subject to the robust stability constraint or the disturbance rejection constraint, respectively. Because the controller design consists of the solution of constrained optimization problems, some concepts from the optimization theory were explained.

References

Ackermann, J., *Robust Control: Systems with Uncertain Physical Parameters*, Springer-Verlag, Berlin, 1993.

Åström, K. and Hägglund T., *PID Controllers: Theory, Design, and Tuning*, Instrumentation Systems, and Automation Society, Research Triangle Park, North Carolina, 1994.

Bernstein, B.S. and Haddad, W.M., LQG control with an H_∞- performance bound: A Riccati equation approach, *IEEE Transactions on Automatic Control*, 34, 3, 293–305, 1994.

Bhattacharyya, S.P., Chapellat, H., and   Keel, L.H., *Robust Control: The Parametric Approach*, Prentice Hall, Englewood Cliffs, New Jersey, 1995.

Boyd, S.T. and Barrat, C.H. *Linear Controller Design: Limit of Performance*, Prentice Hall, Englewood Cliffs, New Jersey, 1991.

Chen, B.-S., Cheng, Y.-M., and Lee, C.-H., A genetic approach to mixed H_2/H_∞ optimal PID control, *IEEE Control Systems Magazine*, 15, 5, 51–60, 1995.

Dixon, L.C.W. and Szegö, G.P., *Towards Global Optimization*, North-Holland Publishing Company, Amsterdam, 1978.

Djaferis, T.E., *Robust Control Design: A Polynomial Approach*, Kluwer Academic, Dordrecht, 1995.

Doyle, J.C., Francis, B.A., and Tannenbaum, A.R., *Feedback Control Theory*, Macmillan Publishing, New York, 1992.

Doyle, J.C. et al., Mixed H_2 and H_∞ performance objectives II: Optimal control, *IEEE Transactions on Automatic Control*, 39, 8, 1575–1587, 1994.

Homaifar, A., Qi, C.X., and Lai, S.H., Constrained optimization via genetic algorithms, *Simulation*, 62, 4, 242–254, 1994.

Joines, J.A. and Houck, C.R., On the use of nonstationary penalty functions to solve nonlinear constrained optimization problems with genetic algorithms, *Proceedings of the Evolutionary Computation Conference, part of the IEEE World Congress on Computational Intelligence*, Orlando, Florida, pp. 579–584, 1994.

Jury, E.I., *Inners and Stability of Dynamic Systems*, John Wiley & Sons, New York, 1974.

Khargonekar, P.P. and Rotea, M.A., Mixed H_2/H_∞ control: A convex optimization approach, *IEEE Transactions on Automatic Control*, 36, 7, 824–837, 1991.

Kim, J.-H. and Myung, H., Evolutionary programming techniques for constrained optimization problems, *IEEE Transactions on Evolutionary Computation*, 1, 2, 129–140, 1997.

Kwaakernaak, H., Robust control and H_∞- optimization — A Tutorial Paper, *Automatica*, 29, 2, 255–273, 1993.

Michalewicz, Z., A survey of constraint handling techniques in evolutionary methods, *Proceedings of the 4th Annual Conference on Evolutionary Programming*, MIT Press, Cambridge, pp. 135–155, 1995.

Michalewicz, Z., *Genetic Algorithms + Data Structure = Evolution Programs*, Springer-Verlag, Berlin, 1996.

Milanese, M., Norton, J., Piet-Lahaniert, H., and Walter, E., *Bounding Approaches to System Identification*, Plenum Publishing, New York, 1996.

Mueller, K., *Entwurf robuster Regelungen*, Teubner Verlag, Stuttgart, 1996.

Schneider, F., Geschlossene Formel zur Berechnung der quadratischen und der zeitbeschwerten quadratischen Regelfläche für kontinuierliche und diskrete Systeme, *Regelungstechnik*, 14, 4, 159–166, 1966.

Smith, R.S. and Dahleh, M., *The Modeling of Uncertainty in Control Systems*, Lecture Notes in Control and Information Sciences, Vol. 192, Springer-Verlag, Berlin, 1994.

Snaizer, M., An exact solution to general SISO mixed H_2/H_∞ problems via convex optimization, *IEEE Transactions on Automatic Control*, 39, 12, 2511–2517, 1994.

Tan, K.C. and Li, Y., L_∞- identification and model reduction using a learning genetic algorithm, *Proceedings of the UKACC International Conference on Control*, IEE Publication no. 427, pp. 1125–1130, 1996.

Vidyasagar, M., *Control System Design: A Factorization Approach*, MIT Press, Cambridge, 1985.

Westcott, J.H., The minimum-moment-of-error-squared criterion: A new performance criterion for servo mechanisms, *IEEE Proceedings*, London, pp. 471–480, 1954.

Zames, G.F., Feedback and optimal sensitivity: Model reference transformations, multiplicative seminorms and approximate inverses, *IEEE Transactions on Automatic Control*, 26, 2, 301–320, 1981.

chapter three

Methods for controller design using genetic algorithms*

3.1 Introduction to controller design using genetic algorithms

The controller design in the previous chapter was formulated as a constrained optimization problem. The challenge is to solve the resulting optimization problem. Classical optimization methods are based on assumptions such as differentiability, convexity of the objective function, as well as of the constraints that must be satisfied. Because these assumptions cannot be satisfied by the constrained optimization on hand, we develop methods based on genetic algorithms. Due to their high potential for global optimization, GAs have received great attention in the area of automatic control with successful applications (Capponetto et al., 1994; Hunt, 1992; Patton and Liu, 1994; Porter II and Passino, 1994; Wang and Kwok, 1994; Krohling, 1997a, 1997b, 1997c, 1998; Krohling et al., 1997a; Krohling and Rey, 2001; Man et al., 1996, 1997; Onnen et al., 1997), and many others to be cited in further chapters. For a more complete list, the reader is refered to Alander (1995). In this chapter, we show how GAs can be used to design optimal robust and optimal disturbance rejection controllers with fixed structure formulated as a constrained optimization problem.

3.2 Design of optimal robust controller with fixed structure

The design of optimal robust controllers was formulated as minimization of the performance index ISE $J_n(k)$ or ITSE $I_n(k)$ subject to the robust stability

* Portions reprinted, with permission, from Krohling, R.A. and Rey, J.P., Design of optimal disturbance rejection PID controllers using genetic algorithms, *IEEE Trans. on Evolutionary Algorithms*, 5, 1, 78–82, Feb. 2001. ©1988/2001 IEEE.

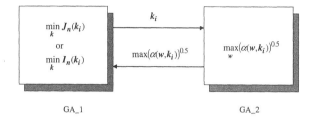

GA_1   GA_2

Figure 3.1 Representation of the method for optimal robust controller design using two GAs.

constraint $\max(\alpha(w,k_i))^{0.5} < 1$. To solve the constrained optimization problem, two genetic algorithms are employed, i.e., GA_1 to minimize the performance index $J_n(k)$ or $I_n(k)$ and GA_2 to maximize the robust stability constraint $\alpha(w,k)$, as depicted in Figure 3.1. Initially, GA_1 is started with the controller parameters within the search domain as specified by the designer. These parameters are then transferred to GA_2, which is initialized with the variable frequency w. GA_2 maximizes the robust stability constraint during a fixed number of generations for each individual of GA_1. Next, if the maximum value of the robust stability constraint is larger than 1, a penalizing value will be assigned to the corresponding individual of GA_1. Individuals of GA_1 that do not satisfy the robust stability constraint will be penalized. In the evaluation of the fitness function of GA_1, individuals with higher fitness values are selected automatically, and those penalized will probably not survive the evolutionary process.

To apply genetic algorithms, it is necessary to define the representation of the individuals, the genetic operators, and the fitness function.

Representation: For GA_1, an individual c consists of the controller parameters (vector k). For GA_2, an individual c consists of only one gene (frequency w). For both GA_1 and GA_2, the individuals were represented as real number and randomly initialized.

Genetic operators: For the implementation of the GAs, we used tournament selection, arithmetical crossover, and mutation (Michalewicz, 1996).

Fitness function: Let the performance index be $J_n(k)$ or $I_n(k)$, then the value of the fitness of each individual of GA_1, $k_i (i = 1,...,\mu_1)$, is determined by the fitness functions, denoted by $F_{1a}(k_i)$ as:

$$F_{1a}(k_i) = -J_n(k_i) - P_{1a}(k_i) \tag{3.1}$$

or

$$F_{1a}(k_i) = -I_n(k_i) - P_{1a}(k_i) \tag{3.2}$$

where μ_1 denotes the population size of GA_1. The penalty function $P_{1a}(k)$ is discussed in the following.

Let the robust stability constraint be max $(\alpha(w,k_i))^{0.5}$. The value of the fitness of each individual of GA_2 $w_j(j = 1,...,\mu_2)$ is determined by the fitness function, denoted by $F_{2a}(w_j)$ as:

$$F_{2a}(w_j) = \alpha(w_j, k_i) \tag{3.3}$$

where μ_2 denotes the population size of GA_2.

The penalty for the individual k_i is calculated by means of the penalty function $P_{1a}(k_i)$ given by:

$$P_{1a}(k_i) = \begin{cases} M_t, & \text{if } k_i \text{ is not stable} \\ M_s\left(\max \alpha(w,k_i)\right), & \text{if max } (\alpha\,(w,k_i))^{0.5} > 1 \\ 0, & \text{if max } (\alpha\,(w,k_i))^{0.5} < 1 \end{cases} \tag{3.4}$$

If the individual k_i does not satisfy the stability test applied to the characteristic equation of the system, then k_i is an unstable individual, and it is penalized with a very large positive constant M_t. Automatically, k_i does not survive the evolutionary process. If k_i satisfies the stability test but not the robust stability constraint, then it is an infeasible individual, and it is penalized with $M_s(\max \alpha(w,k_i))$, where M_s is a positive constant to be adjusted. Otherwise, the individual k_i is feasible and is not penalized.

3.2.1 Design method

The method for design of optimal robust controller with fixed structure can be summarized as follows (Krohling, 1998):

- Given the plant with transfer function $G_0(s)$, the controller with fixed structure and transfer function $C(s,k)$, and the weighting function $W_m(s)$, determine the error signal $E(s)$ and the robust stability constraint $\alpha(w,k)$.
- Specify the lower and upper bounds of the controller parameters.
- Set up GA_1 and GA_2 parameters: crossover probability, mutation probability, population size, and maximum number of generations.

It is convenient to describe the method in the form of an algorithm.

Step 1: Initialize the populations of GA_1 $k_i(i = 1,...,\mu_1)$ and GA_2 $w_j(j = 1,...,\mu_2)$, and set the generation number of GA_1 to $g_1 = 1$, where g_1 denotes the number of generations for GA_1.

Step 2: For each individual k_i of the GA_1 population, calculate the maximum value of $\alpha(w,k_i)$ using GA_2. If no individuals of the GA_1 satisfy the constraint max $(\alpha(w,k_i))^{0.5} < 1$, then a feasible solution is assumed to be nonexistent, and the algorithm stops. In this case, a new controller structure has to be assumed.

Step 3: For each individual k_i of GA_1, the penalty value is calculated by using Equation (3.4), and the fitness value is calculated by using Equations (3.1) or (3.2).

Step 4: Select individuals using tournament selection, and apply genetic operators (crossover and mutation) to the individuals of GA_1.

Step 5: For each individual k_i of the GA_1, calculate max $(\alpha(w,k_i))^{0.5}$ using GA_2 as follows:

Substep a: Initialize the gene of each individual $w_j(j = 1,...,\mu_2)$ in the population and set the generation number to $g_2 = 1$, where g_2 indicates the number of generations for GA_2.

Substep b: Evaluate the fitness of each individual by using Equation (3.3).

Substep c: Select individuals using tournament selection, and apply genetic operators (crossover and mutation).

Substep d: If the maximum number of generations of GA_2 is reached, stop and return the fitness of the best individual max $(\alpha(w,k_i))$ to GA_1, otherwise set $g_2 = g_2 + 1$, and go to Substep b.

Step 6: If the maximum number of generations of GA_1 is reached, stop. Otherwise, set $g_1 = g_1 + 1$ and go to Step 3.

The best individual k^* (vector of the controller parameters) represents the optimal solution of the optimization problem. The best individual is first checked to see whether the condition for robust stability is fulfilled, i.e., max $(\alpha(w,k_i^*))^{0.5} < 1$. If this is not the case, the optimization must be repeated, because this solution represents an unfeasible individual. Otherwise, the controller will be tested. By means of simulation studies, the control system is excited with a unit step, and the performance can be evaluated for the nominal plant as well as for the plant subject to uncertainty.

3.2.2   *Design example*

To illustrate the method, a detailed design example is presented. Consider the control system shown in Figure 3.2.

The plant is described by the following transfer function (Lo Bianco and Piazzi, 1997):

$$G_0(s) = \frac{1.8}{s^2(s+2)} \qquad (3.5)$$

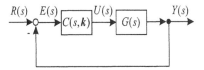

Figure 3.2 Control system with uncertain plant.

The controller $C(s,k)$ is described by the following transfer function (Lo Bianco and Piazzi, 1997):

$$C(s,k) = k_1 \frac{s^2 + 2k_4 k_5 s + k_5^2}{(s + k_2)(s + k_3)} \qquad (3.6)$$

The vector k of the controller parameter is given by $k = [k_1, k_2, k_3, k_4, k_5]^T$.
The multiplicative uncertainty $W_m(s)$ is given by (Lo Bianco and Piazzi, 1997):

$$W_m(s) = \frac{0.1}{s^2 + 0.1s + 10} \qquad (3.7)$$

The error signal $E(s)$, assuming the input signal is a unit step, is evaluated as follows:

$$E(s) = \frac{1}{1 + C(s,k)G_0(s)} \cdot R(s)$$

By introducing $C(s,k)$, $G_0(s)$ and $R(s)$ follows:

$$E(s) = \frac{d_0 s^4 + d_1 s^3 + d_2 s^2 + d_3 s + d_4}{a_0 s^5 + a_1 s^4 + a_2 s^3 + a_3 s^2 + a_4 s + a_5} \qquad (3.8)$$

The squared error $J_5(k)$ is given by (Schneider, 1966; Westcott, 1954):

$$J_5(k) = \frac{Z_1}{N_1} \qquad (3.9)$$

where

$$Z_1 = a_2 a_3 a_4 a_5 d_0^2 - a_1 a_4^2 a_5 d_0^2 - a_2^2 a_5^2 d_0^2 + a_0 a_4 a_5^2 d_0^2 + a_0 a_3 a_4 a_5 d_1^2 - a_0 a_2 a_5^2 d_1^2 -$$

$$- 2a_0 a_3 a_4 a_5 d_0 d_2 + 2a_0 a_2 a_5^2 d_0 d_2 + a_0 a_1 a_4 a_5 d_2^2 - a_0^2 a_5^2 d_2^2 - 2a_0 a_1 a_4 a_5 d_1 d_3 +$$

$$+ 2a_0^2 a_5^2 d_1 d_3 + a_0 a_1 a_2 a_5 d_3^2 - + a_0^2 a_3 a_5 d_3^2 + 2a_0 a_1 a_4 a_5 d_0 d_4 - 2a_0^2 a_5^2 d_0 d_4 - 2a_0 a_1 a_2 a_5 d_2 d_4 +$$

$$+ 2a_0^2 a_3 a_5 d_2 d_4 + a_0 a_1 a_2 a_3 d_4^2 - a_0^2 a_3^2 d_4^2 - a_0 a_1^2 a_4 d_4^2 + a_0^2 a_1 a_5 d_4^2$$

$$N_1 = 2a_0a_5\left(a_1a_2a_3a_4 - a_0a_3^2a_4 - a_1^2a_4^2 - a_1a_2^2a_5 + a_0a_2a_3a_5 + 2a_0a_1a_4a_5 - a_0^2a_5^2\right)$$

The time-weighted squared error $I_5(k)$ is given (Schneider, 1966; Westcott, 1954):

$$I_5(k) = \frac{Z_2}{N_2} + \frac{Z_3}{N_3} \tag{3.10}$$

where

$$N_3 = 4a_0a_5^2\left(a_1a_2a_3a_4 - a_0a_3^2a_4 - a_1^2a_4^2 - a_1a_2^2a_5 + a_0a_2a_3a_5 + 2a_0a_1a_4a_5 - a_0^2a_5^2\right)^2$$

$$Z_3 = Z_4(a_1a_2a_3a_4^2 - a_0a_3^2a_4^2 - a_1^2a_4^3 + 2a_1a_2a_3^2a_5 - 2a_0a_3^3a_5 + a_1a_2^2a_4a_5 + a_0a_2a_3a_4a_5 - \\ - 6a_0a_1a_4^2a_5 - 8a_1^2a_2a_5^2 - 2a_0a_2^2a_5^2 + 8a_0a_1a_3a_5^2 + 7a_0^2a_4a_5^2)$$

$$N_2 = 4a_0a_5\left(a_1a_2a_3a_4 - a_0a_3^2a_4 - a_1^2a_4^2 - a_1a_2^2a_5 + a_0a_2a_3a_5 + 2a_0a_1a_4a_5 - a_0^2a_5^2\right)$$

$$Z_2 = -(a_2a_3a_4^2d_0^2 - a_1a_4^3d_0^2 + 2a_2a_3^2a_5d_0^2 + a_2^2a_4a_5d_0^2 - 3a_0a_4^2a_5d_0^2 - 8a_1a_2a_5^2d_0^2 + 2a_0a_3a_5^2d_0^2 \\ +2a_0a_3a_4a_5d_0d_1 - 2a_0a_2a_5^2d_0d_1 + a_0a_3a_4^2d_1^2 + 2a_0a_3^2a_5d_1^2 + a_0a_2a_4a_5d_1^2 - 4a_0a_1a_5^2d_1^2 - \\ -2a_0a_3a_4^2d_0d_1 - 4a_0a_3^2a_5d_0d_2 - 2a_0a_2a_4a_5d_0d_2 + 8a_0a_2a_5^2d_0d_2 + 2a_0a_1a_4a_5d_1d_2 - \\ -2a_0^2a_5^2d_1d_2 + a_0a_1a_4^2d_2^2 + 2a_0a_1a_3a_5d_2^2 + 3a_0^2a_4a_5d_2^2 - 6a_1a_2a_4a_5d_0d_3 + 6a_0^2a_5^2d_0d_3 - \\ -2a_0a_1a_4^2d_1d_3 - 4a_0a_1a_3a_5d_1d_3 - 6a_0^2a_4a_5d_1d_3 + 2a_0a_1a_2a_5d_2d_3 - 2a_0^2a_3a_5d_2d_3 + \\ +a_0a_1a_2a_4d_3^2 - a_0^2a_3a_4d_3^2 + 4a_0a_1^2a_5d_3^2 + 2a_0^2a_2a_5d_3^2 + 2a_0a_1a_4^2d_0d_4 + 4a_0a_1a_3a_5d_0d_4 + \\ +6a_0^2a_4a_5d_0d_4 - 6a_0a_1a_2a_5d_1d_4 + 6a_0^2a_3a_5d_1d_4 - 2a_0a_1a_2a_4d_2d_4 + 2a_0^2a_3a_4d_2d_4 - \\ -8a_0a_1^2a_5d_2d_4 - 4a_0^2a_2a_5d_2d_4 + 2a_0a_1a_2a_3d_4 - 2a_0^2a_3^2d_3d_4 - 2a_0a_1^2a_4d_3d_4 + \\ +2a_0^2a_1a_5d_3d_4 + 3a_0a_1a_2^2d_4^2 + 2a_0a_1^2a_3d_4^2 - a_0^2a_2a_3d_4^2 - 9a_0^2a_1a_4d_4^2 + 5a_0^3a_5d_4^2)$$

The robust stability constraint $(\alpha(w,k))^{0.5}$ is calculated using the software tool Mathematica (Wolfram, 1988):

$$\alpha(w,k) = \frac{\alpha_z(w,k)}{\alpha_n(w,k)} \tag{3.11}$$

where

$$\alpha_z(w,k) = 0.0324k_1^2k_5^4 + (-0.0648k_1^2k_5^2 + 0.1296k_1^2k_4^2k_5^2)w^2 + (0.0324k_1^2)w^4$$

$$a_n(w,k) = [100 - 19.99w^2 + w^4][3.24k_1^2k_5^4 + (6.48k_1^2k_5^2 - 7.2k_1k_2k_3k_5^2 + 12.96k_1^2k_4^2k_5^2)w^2 +$$

$$+ (3.24k_1^2 + 7.2k_1k_2k_3 + 4k_2^2k_3^2 - 14.4k_1k_2k_4k_5 - 14.4k_1k_3k_4k_5 - 7.2k_1k_2k_3k_4k_5 +$$

$$+ 7.2k_1k_5^2 + 3.6k_1k_2k_5^2 + 3.6k_1k_3k_5^2)w^4 + (-7.2k_1 - 3.6k_1k_2 + 4k_2^2 - 3.6k_1k_3 + 4k_3^2 +$$

$$+ k_2^2k_3^2 + 7.2k_1k_4k_5)w^6 + (4 + k_2^2 - k_3^2)w^8 + w^{10}]$$

The controller parameter was searched in the following bounds (Lo Bianco and Piazzi, 1997):

$$k_1 = [1,\ 1000];\ k_2 = [1,\ 100];\ k_3 = [1,\ 100];\ k_4 = [0,\ 1];\ k_5 = [0.1,\ 100]$$

So, the method using GA can be applied to the design of the controller.

The GA parameters were kept constant for all simulations with crossover probability $p_{c1} = p_{c2} = 0.35$, mutation probability $p_{m1} = p_{m2} = 0.02$, population size for GA_1 $\mu_1 = 100$, population size for GA_2 $\mu_2 = 50$, penalty constant $M_t = 1,000,000$, penalty constant $M_s = 100$, maximum number of generations for GA_1 $g_{1max} = 100$, and maximum number of generations for GA_2 $g_{2max} = 50$. The values for crossover probability and mutation probability follow standard implementations in the literature (e.g., Michalewicz, 1996).

The convergence of the minimization of the ISE performance index $J_5(k)$ subject to the disturbance rejection constraint max $(\alpha(w,k))^{0.5} < 1$ by using GA_1 for the best individual during the first 50 generations is shown in Figure 3.3. The ordinate does not begin at zero for a clearer presentation of

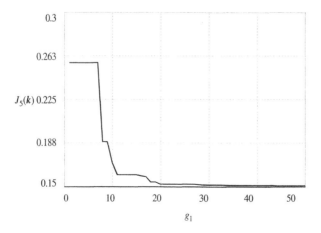

Figure 3.3 Convergence of the minimization of the ISE performance index $J_5(k)$ subject to the robust stability constraint max $(\alpha(w,k))^{0.5} < 1$.

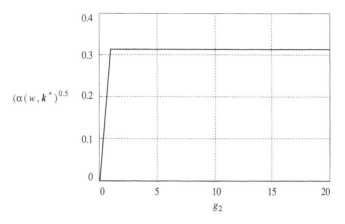

Figure 3.4 Calculation of the maximum value of to the robust stability constraint max $(\alpha(w,k^*))^{0.5}$.

the results. The minimum value $J_5(k^*) = 0.1507$ is achieved in 44 generations, and the corresponding best individual, i.e., the vector of controller parameters, yields $k^* = [1000.0, 14.518, 14.519, 1.0, 0.542]^T$.

   The calculation of the maximum value of the robust stability constraint for the optimal vector of controller parameters k^* by using GA_2 is shown in Figure 3.4. The maximum value is $\alpha(w,k^*)^{0.5} = 0.313$. Because this value is smaller than 1, k^* represents a feasible individual. Therefore, the condition for robust stability is satisfied.

   The convergence of the minimization of the ITSE performance index $I_5(k)$ subject to the disturbance rejection constraint, max $(\alpha(w,k))^{0.5} < 1$ by using GA_1 for the best individual during the first 50 generations, is shown in Figure 3.5. The minimum value $I_5(k^*) = 0.0252$ is achieved in 35 generations, and the corresponding best individual, i.e., the vector of controller parameters, yields $k^* = [1000.0, 15.328, 15.073, 1.0, 0.893]^T$.

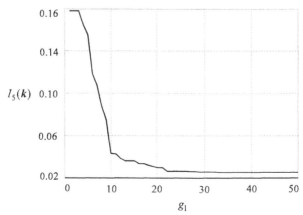

Figure 3.5 Convergence of the minimization of the ITSE performance index $J_5(k)$ subject to the robust stability constraint max $(\alpha(w,k))^{0.5} < 1$.

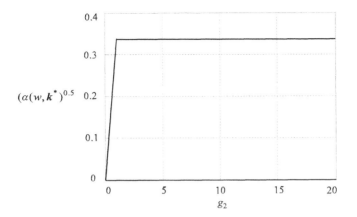

Figure 3.6 Calculation of the maximum value of to the robust stability constraint max $(\alpha(w,k^*))^{0.5}$.

   The calculation of the maximum value of the robust stability constraint for the optimal vector of controller parameters k^* by using GA_2 is shown in Figure 3.6. The maximum value is $(\alpha(w, k^*))^{0.5} = 0.341$. Because this value is smaller than 1, it means that k^* represents a feasible individual. Therefore, the condition for robust stability is satisfied.
   The performance of the control system shown in Figure 3.1, using a controller design based on the proposed method, is tested by closed-loop step response for two cases: nominal plant and plant with uncertainty. In order to verify the suitability of the controller tuning, the plant is excited with a unit step. The tolerance for the controlled variable is ±2% of the set point amplitude. The tracking behavior of the control system, which was determined by the minimization of the ISE performance index $J_5(k)$, is shown in Figure 3.7 for the nominal plant $G_0(s)$ and for the plant with uncertainty $G(s)$.

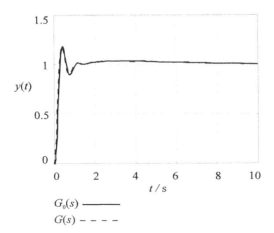

Figure 3.7 Unit step response for the plant with uncertainty [determined by the minimization of the ISE performance index $J_5(k)$].

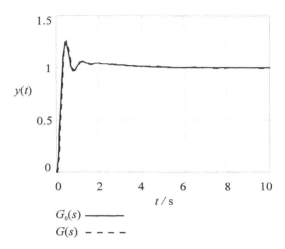

Figure 3.8 Unit step response for the plant with uncertainty [determined by the minimization of the ITSE performance index $I_5(k)$].

The tracking behavior of the control system, which was determined by the minimization of the ITSE performance index $I_5(k)$, is shown in Figure 3.8 for the nominal plant $G_0(s)$ and for the plant with uncertainty $G(s)$.

The closed-loop step response of the control system, which was designed by minimizing the performance index ISE, presents about 17% of overshoot for a tolerance of ±2% of the set-point amplitude, and the settling time is about 5 s. Figure 3.7 shows that the closed-loop step response for the plant with uncertainty presents almost no significant change when compared to the nominal plant. Therefore, the influence of the uncertainty acting on the plant is very small, and this confirms the suitability of the method.

The closed-loop step response of the control system, which was designed by minimizing the performance index ITSE, presents about 22% of overshoot for a tolerance of ±2% of the set-point amplitude, and the settling time is about 2.5 s. Figure 3.8 shows that the closed-loop step response for the plant with uncertainty deviates little from that of the nominal plant. Therefore, the influence of the uncertainty acting on the plant output is very small. The results obtained clearly show the effectiveness of the proposed method in the design of an optimal robust controller with fixed structure. If not optimal, at least a good solution was found.

3.3   *Design of optimal disturbance rejection controller with fixed structure*

The design of optimal disturbance rejection controllers was formulated as minimization of the performance index ISE $J_n(k)$ or the ITSE $I_n k$ subject to the disturbance rejection constraint max $(\beta (w, k_i))^{0.5} < \gamma$. For the solution of the constrained optimization problem, two genetic algorithms are employed,

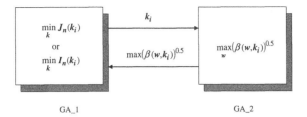

GA_1   GA_2

Figure 3.9 Representation of the method for optimal disturbance rejection controller design using two GAs. (From Krohling, R.A. and Rey, J.P., *IEEE Trans. on Evolutionary Algorithms*, 5, 1, 78–82, Feb. 2001, ©1988/2001 IEEE. With permission.)

i.e., GA_1 to minimize the performance index $J_n(k)$ or $I_n(k)$ and GA_2 to maximize the disturbance rejection constraint $(\beta(w, k_i))^{0.5}$ as depicted in Figure 3.9. Initially, GA_1 is started with the controller parameters within the search domain as specified by the designer. These parameters are then transferred to GA_2, which is initialized with the variable frequency w. GA_2 maximizes the disturbance rejection constraint during a fixed number of generations for each individual of GA_1. Next, if the maximum value of the disturbance rejection constraint is larger than γ, a penalizing value will be assigned to the corresponding individual of GA_1. Individuals of GA_1 that do not satisfy the disturbance rejection constraint will be penalized. In the evaluation of the fitness function of GA_1, individuals with higher fitness values are selected automatically, and those penalized will not survive the evolutionary process. For the implementation of the GAs, we used tournament selection, arithmetic crossover, and mutation (Michalewicz, 1996).

The representation of the individuals and the genetic operators used here are the same ones used in the last section. Therefore, only the fitness function needs to be redefined.

Fitness Function

Let the performance index be $J_n(k)$ or $I_n(k)$, then the value of the fitness of each individual of GA_1 $k_i(i = 1,\dots,\mu_1)$ is determined by the fitness functions, denoted by $F_{1b}(k_i)$:

$$F_{1b}(k_i) = -J_n(k_i) - P_{1b}(k_i) \tag{3.12}$$

or

$$F_{1b}(k_i) = -I_n(k_i) - P_{1b}(k_i) \tag{3.13}$$

Where μ_1 denotes the population size of GA_1. The penalty function $P_{1b}(k)$ is discussed in the following.

Let the disturbance rejection constraint be max $(\beta(w,k_i))^{0.5}$. The value of the fitness of each individual of GA_2 $w_j(j = 1,...,\mu_2)$ is determined by the fitness function, denoted by $F_{2b}(w_j)$:

$$F_{2b}(w_j) = \beta(w_j,k_i) \tag{3.14}$$

where μ_2 denotes the population size of GA_2.

Let the disturbance rejection constraint be max $(\beta(w,k_i))^{0.5}$. The value of the fitness of each individual of GA_2 $w_j(j = 1,...,\mu_2)$ is determined by the fitness function, denoted by $F_{2b}(w_j)$.

The penalty for the individual k_i is calculated by means of the penalty function $P_{1b}(k_i)$ given by:

$$P_{1b}(k_i) = \begin{cases} M_t, & \text{if } k_i \text{ is not stable} \\ M_s\left(\max \beta\,(w,k_i)\right), & \text{if } \max\,(\beta\,(w,k_i))^{0.5} > \gamma \\ 0, & \text{if } \max\,(\beta\,(w,k_i))^{0.5} < \gamma \end{cases} \tag{3.15}$$

If the individual k_i does not satisfy the stability test applied to the characteristic equation of the system, then k_i is an unstable individual, and it is penalized with a very large positive constant M_t. Automatically, k_i does not survive the evolutionary process. If k_i satisfies the stability test but not the disturbance rejection constraint, then it is an infeasible individual, and it is penalized with M_s (max $\beta(w,k_i)$), where M_s is a positive constant to be adjusted. Otherwise, the individual k_i is feasible and is not penalized.

3.3.1 Design method

The method for design of optimal disturbance rejection controller with fixed structure can be summarized as follows (Krohling and Rey, 2001):

- Given the plant with transfer function $G_0(s)$, the controller with fixed structure and transfer function $C(s,k)$, and the weighting function $W_d(s)$, determine the error signal $E(s)$ and the disturbance rejection constraint $\beta(w,k)$.
- Specify the lower and upper bounds of the controller parameters.
- Set up GA_1 and GA_2 parameters: crossover probability, mutation probability, population size, and maximum number of generations.

It is more convenient to describe the method in the form of an algorithm.

Step 1: Initialize the populations of GA_1 $k_i(i = 1,...,\mu_1)$ and GA_2 w_j ($j = 1,...,\mu_2$), and set the generation number of GA_1 to $g_1 = 1$, where g_1 denotes the number of generations for GA_1.

Step 2: For each individual k_i of the GA_1 population, calculate the maximum value of $\beta(w,k_i)$ using GA_2. If no individuals of the GA_1 satisfy the constraint max $(\beta(w,k_i))^{0.5} < \gamma$, then a feasible solution is assumed to be nonexistent, and the algorithm stops. In this case, a new controller structure has to be assumed.

Step 3: For each individual k_i of GA_1 is calculated the penalty value by using Equation (3.15) and the fitness value by using Equations (3.12) or (3.13).

Step 4: Select individuals using tournament selection and apply genetic operators (crossover and mutation) to the individuals of GA_1.

Step 5: For each individual k_i of the GA_1, calculate max $(\beta(w, k_i))^{0.5}$ using GA_2 as follows:

Substep a: Initialize the gene of each individual $w_j(j = 1,...,\mu_2)$ in the population, and set the generation number to $g_2 = 1$, where g_2 indicates the number of generations for GA_2.

Substep b: Evaluate the fitness of each individual by using Equation (3.14).

Substep c: Select individuals using tournament selection, and apply genetic operators (crossover and mutation).

Substep d: If the maximum number of generations of GA_2 is reached, stop, and return the fitness of the best individual max $(\beta(w,k_i))$ to GA_1. Otherwise, set $g_2 = g_2 + 1$, and go to Substep b.

Step 6: If the maximum number of generations of GA_1 is reached, stop. Otherwise set $g_1 = g_1 + 1$ and go to Step 3.

The best individual k^* (vector of the controller parameters) represents the optimal solution of the optimization problem. First, the best individual k^* is checked, whether or not the condition for disturbance rejection is fulfilled, i.e., max $(\beta(w,k_i))^{0.5} < \gamma$. If this is not the case, the optimization must be repeated, because this solution represents an unfeasible individual. Otherwise, the controller will be tested. By means of simulation studies, the control system is excited with a unit step, and the performance can be evaluated for the nominal plant as well as for the plant subject to disturbance.

3.3.2 Design example

To illustrate the method, a detailed design example is presented. Consider the control system shown in Figure 3.10.

The plant, a servomotor, is described by the following transfer function (Chen et al., 1995):

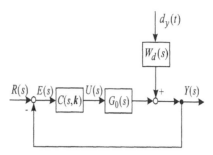

Figure 3.10 Control system with disturbance acting on the plant. (From Krohling, R.A. and Rey, J.P., *IEEE Trans. on Evolutionary Algorithms*, 5, 1, 78–82, Feb. 2001, ©1988/2001 IEEE. With permission.)

$$G_0(s) = \frac{0.8}{s(0.5s+1)} \tag{3.16}$$

The weighting function $W_d(s)$ is given by the following (Chen et al., 1995):

$$W_d(s) = \frac{1}{s+1} \tag{3.17}$$

The disturbance is considered to be $d_y(t) = 0.1\sin t$, and the disturbance attenuation level specified is $\gamma = 0.1$.

The controller $C(s,k)$ is described by the following transfer function (Chen et al., 1995):

$$C(s,k) = k_1 + \frac{k_2}{s} + k_3 s \tag{3.18}$$

The vector k of the controller parameter is given by:

$$k = \left[k_1, k_2, k_3\right]^T = \left[k_p, k_i, k_d\right]^T.$$

Because the plant, as described by Equation (3.16), already contains an integral action, a controller with integral action (k_2) to control this plant is not necessary.

The error signal $E(s)$, assuming the input signal is a unit step, is evaluated as follows:

$$E(s) = \frac{1}{1 + C(s,k)G_0(s)} \cdot R(s)$$

By introducing $C(s,k)$, $G_0(s)$ and $R(s)$ follow:

$$E(s) = \frac{d_0 s^2 + d_1 s}{a_0 s^3 + a_1 s^2 + a_2 s + a_3} \tag{3.19}$$

where $d_0 = 0.5$, $d_1 = 1$, $d_2 = 0$, $a_0 = 0.5$, $a_1 = 1 + 0.8k_3$, $a_2 = 0.8k_1$, $a_3 = 0.8k_2$.

The squared error $J_3(k)$ is given by (Schneider, 1966; Westcott, 1954):

$$J_3(k) = \frac{a_2 a_3 d_0^2 + a_0 a_3 d_1^2 - 2a_0 a_3 d_0 d_2 + a_0 a_1 d_2^2}{2a_0 a_3 (a_1 a_2 - a_0 a_3)} \tag{3.20}$$

The time-weighted squared error $I_3(k)$ is given by (Schneider, 1966; Westcott, 1954):

$$I_3(k) = -\frac{a_2^2 d_0^2 + 2a_1 a_3 d_0^2 + 2a_0 a_3 d_0 d_1 + a_0 a_2 d_1^2 - 2a_0 a_1 d_1 d_2 + 3a_0^2 d_2^2}{4a_0 a_3 (a_1 a_2 - a_0 a_3)} +$$

$$+\frac{\left(a_2 a_3 d_0^2 + a_0 a_3 d_1^2 + 2a_0 a_3 d_0 d_2 + a_0 a_1 d_2^2\right)\left(a_1 a_2^2 + 2a_1^2 a_3 + a_0 a_2 a_3\right)}{4a_0 a_3^2 (a_1 a_2 - a_0 a_3)^2} \tag{3.21}$$

The disturbance rejection constraint $\beta(w,k)^{0.5}$ is calculated using the software tool Mathematica (Wolfram, 1988):

$$\beta(w,k) = \frac{\beta_z(w,k)}{\beta_n(w,k)} \tag{3.22}$$

where

$$\beta_z(w,k) = w^4 + 0.25w^6$$

$$\beta_n(w,k) = 0.64k_2^2 + (-1.6k_2 - 1.28k_3 k_2 + 0.64k_2^2 + 0.64k_1^2)w^2 + (1 + 1.6k_3 - 1.6k_i -$$

$$-1.28k_3 k_2 - 0.8k_1 + 0.64k_1^2)w^4 + (1.25 + 1.6k_3 + 0.64k_3^2 - 0.8k_1)w^6 + 0.25w^8$$

The controller parameter was searched in the following bounds (Chen et al., 1995):

$$k_1 = [0,\ 30];\ k_2 = [0,\ 30];\ k_3 = [0,\ 30]$$

The method GA has been applied to the design of the PID controller.
The GA parameters were kept constant for all the simulations with crossover probability $p_{c1} = p_{c2} = 0.35$, mutation probability $p_{m1} = p_{m2} = 0.02$,

population size for GA_1 $\mu_1 = 100$, population size for GA_2 $\mu_2 = 50$, penalty constant $M_t = 1,000,000$, penalty constant $M_s = 100$, maximum number of generations for GA_1 $g_{1max} = 100$, and maximum number of generations for GA_2 $g_{2max} = 50$. The values for crossover probability and mutation probability follow standard implementations in the literature (e.g., Michalewicz, 1996).

The convergence of the minimization of the ISE performance index $J_3(k)$ subject to the disturbance rejection constraint max $(\beta(w,k^*))^{0.5} < \gamma$ by using GA_1 for the best individual during the first 20 generations, is shown in Figure 3.11. The minimum value $J_3(k^*) = 0.01083$ is achieved in 12 generations, and the corresponding best individual, i.e., the vector of controller parameters, yields $k^* = [29.988, 0.184, 30.0]^T$.

The calculation of the maximum value of the disturbance rejection constraint for the optimal vector of controller parameters k^* by using GA_2 is shown in Figure 3.12. The maximum value is $(\beta(w, k^*))^{0.5} = 0.02376$. Because this value is smaller than γ, it means that k^* represents a feasible individual. Therefore, the condition for disturbance rejection is satisfied.

The convergence of the minimization of the ITSE performance index $I_3(k)$, subject to the disturbance rejection constraint max $(\beta(w, k^*))^{0.5} < \gamma$, by using GA_1 for the best individual during the first 20 generations, is shown in Figure 3.13. The minimum value $I_3(k^*) = 0.000668$ is achieved in 11 generations, and the corresponding best individual, i.e., the vector of controller parameters, yields $k^* = [29.992, 0.00001, 28.3819]^T$. The result reveals that optimization using the GA yields $k_2 = k_i = 0$. This confirms the suitability of GA as an optimization method, because there is no need for an integral action to control the plant as explained previously.

The calculation of the maximum value of the disturbance rejection constraint for the optimal vector of controller parameters k^* by using GA_2 as shown in Figure 3.14. The maximum value is $(\beta(w, k^*))^{0.5} = 0.02460$. Because

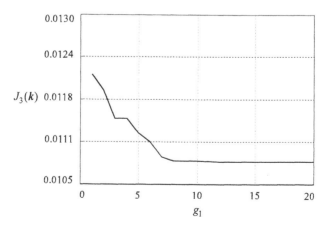

Figure 3.11 Convergene of the minimization of the ISE performance index $J_3(k)$ subject to the disturbance rejection constraint max $(\beta(w,k))^{0.5} < \gamma$.

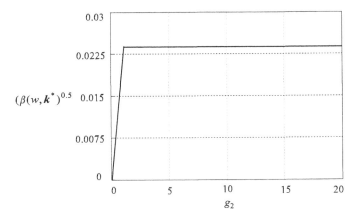

Figure 3.12 Calculation of the maximum value of the disturbance rejection constraint max $(\beta(w,k^*))^{0.5}$.

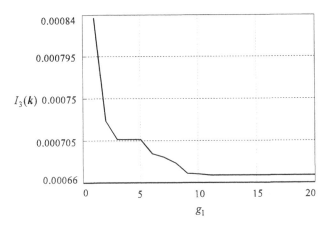

Figure 3.13 Convergence of the minimization of the ITSE performance index $I_3(k)$ subject to the disturbance rejection constraint max $(\beta(w,k))^{0.5} < \gamma$. (From Krohling, R.A. and Rey, J.P., *IEEE Trans. on Evolutionary Algorithms*, 5, 1, 78–82, Feb. 2001, ©1988/2001 IEEE. With permission.)

this value is smaller than γ, it means that k^* represents a feasible individual. Therefore, the condition for disturbance rejection is satisfied. The results obtained using the ITSE performance index show an improved performance as compared to those employing the ISE performance index (Chen et al., 1995).

The performance of the control system in Figure 3.10, using a controller design based on the proposed method, is tested by closed-loop step response for two cases: without disturbance and with disturbance acting on the output of the plant. In order to verify the suitability of the controller tuning, the plant is excited with a unit step. The tolerance for the controlled variable is

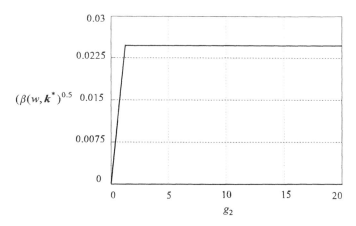

Figure 3.14 Calculation of the maximum value of the disturbance rejection constraint max $(\beta(w,k^*))^{0.5}$. (From Krohling, R.A. and Rey, J.P., *IEEE Trans. on Evolutionary Algorithms*, 5, 1, 78–82, Feb. 2001, ©1988/2001 IEEE. With permission.)

±2% of the set point amplitude. The tracking behavior of the control system, which was determined by the minimization of the ISE performance index $J_3(k)$ is shown in Figure 3.15 for $d_y(t) = 0$ and for a disturbance acting on the plant $d_y(t) = 0.1\sin t$ (Chen et al., 1995).

The step response of the control system with the controller parameters (vector k^*), which was obtained by minimizing the ISE performance index $J_3(k)$, for $d_y(t) = 0$ and for a unit step disturbance $d_y(t) = 1(t)$, is shown in Figure 3.16.

The step response of the control system, which was determined by the minimization of the time-weighted squared error $I_3(k)$, is shown in Figure 3.17 for $d_y(t) = 0$ and for load disturbance $d_y(t) = 0.1\sin t$.

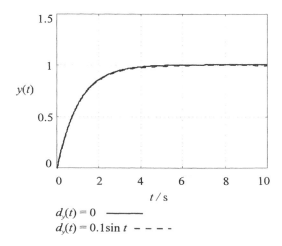

$d_y(t) = 0$ ———
$d_y(t) = 0.1\sin t$ – – – –

Figure 3.15 Unit step response with a sinusoidal disturbance.

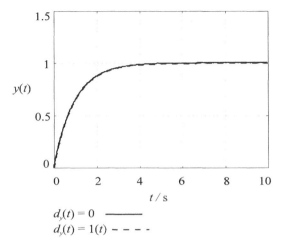

Figure 3.16 Unit step response with a unit step disturbance [determined by the minimization of the ISE performance index $J_3(k)$].

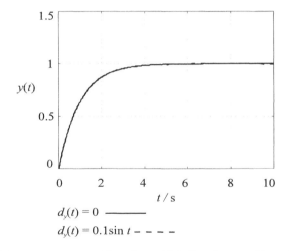

Figure 3.17 Unit step response with a sinusoidal disturbance. (From Krohling, R.A. and Rey, J.P., *IEEE Trans. on Evolutionary Algorithms*, 5, 1, 78–82, Feb. 2001, ©1988/2001 IEEE. With permission.)

The step response of the control system with the controller parameters (vector k^*), which was obtained by minimizing the ITSE performance index $I_3(k)$, for $d_y(t) = 0$ and unit step disturbance $d_y(t) = 1(t)$ is shown in Figure 3.18.

For the control system that was designed by minimizing the performance index ISE, the closed-loop step responses present no overshoot, and the settling time is about 4 sec for a tolerance of ±2% of the set-point amplitude. It can be observed from Figures 3.15 and 3.16 that in closed-loop step response for the plant with sinusoidal disturbance $d_y(t) = 0.1\sin t$ or unit step disturbance $d_y(t) = 1(t)$, almost no significant change occurs when compared

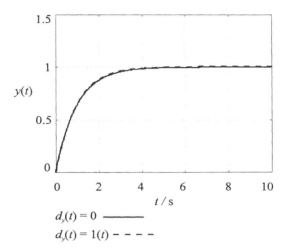

$$d_y(t) = 0 \quad \text{———}$$
$$d_y(t) = 1(t) \quad \text{– – – –}$$

Figure 3.18 Unit step response with a unit step disturbance [determined by the minimization of the time-weighted squared error $I_3(k)$]. (From Krohling, R.A. and Rey, J.P., *IEEE Trans. on Evolutionary Algorithms*, 5, 1, 78–82, Feb. 2001, ©1988/2001 IEEE. With permission.)

to the nominal case (i.e., without disturbance). Therefore, the influence of the disturbance acting on the plant output is small, and this confirms the suitability of the method.

For the control system that was designed by minimizing the performance index ITSE, the closed-loop step responses present no overshoot, and the settling time is about 4 sec for a tolerance of ±2% of the set-point amplitude. It can be observed from Figures 3.17 and 3.18 that the closed-loop step response for the plant with either sinusoidal disturbance $d_y(t) = 0.1\sin t$ or unit step disturbance $d_y(t) = 1(t)$ presents almost no difference compared to the nominal case (i.e., without disturbance). Therefore, the disturbance acting on the plant output has little influence on the step response.

The good step response of the control system, with disturbance acting on the plant for both cases, is due to the small value of the disturbance rejection constraint. In this way, the influence of the disturbance can be limited significantly. The results obtained clearly show the effectiveness of the proposed method in the design of an optimal disturbance rejection controller with fixed structure for the case of a PID.

The obtained results demonstrate the suitability of the developed methods for the design of optimal disturbance rejection controllers. If not optimal, at least a good solution was found.

3.4   *Evaluation of the methods*

Because the developed methods are similar, they have the same characteristics concerning generality, efficiency, simplicity, and practical relevance.

- *Generality*: It is not necessary to make assumptions about the objective function, e.g., differentiability or convexity.
- *Efficiency*: The use of GA with real representation to solve a nonlinear optimization problem has proved to be very suitable. The examples showed that GA find a solution in few generations. Beyond that, there is no problem with numeric precision.
- *Simplicity*: Because GA are easily implemented, the practice application of the developed procedure is connected only with a small expenditure. For each optimization problem, it is only necessary to define the fitness functions and the penalty function.
- *Practical relevance*: The methods enable the control engineer to employ the theory to design more robust controllers, like PID controllers, which have high acceptance in industry applications.

3.5 Conclusions

In this chapter, methods to design optimal robust controllers and optimal disturbance rejection controllers, both with fixed structure, were presented. The basic idea of the methods is that of seeking solutions of constrained optimization problems by using genetic algorithms. The method consists of two genetic algorithms. One genetic algorithm minimizes the performance index (the integral of the squared error or the integral of the time-weighted squared error), and the other maximizes the robust stability constraint or the disturbance rejection constraint. The entire design process was described in terms of algorithms. The methods for controller design were illustrated by two examples. The advantages of these methods over known methods were shown.

References

Alander, J.T., *An Indexed Bibliography of Genetic Algorithms in Control*, Report 94-1, Department of Information Technology and Production Economics, University of Vaasa, Finland, 1995.

Capponetto, R. et al., Genetic algorithms for controller order reduction, *Proceedings of the 1st IEEE Conference on Evolutionary Computation*, Orlando, Florida, Vol. 2, pp. 724–729, 1994.

Chen, B.-S., Cheng, Y.-M., and Lee, C.-H., A genetic approach to mixed H_2/H_∞ optimal PID control, *IEEE Control Systems Magazine*, 15, 5, 51–60, 1995.

Hunt, K.J., Polynomial LQG and H-infinity controller synthesis — A genetic algorithm solution, *Proceedings of the 31st IEEE Conference on Decision and Control*, Tucson, Arizona, pp. 3604–3609, 1992.

Krohling, R.A. Synthesis of PID controller using genetic algorithms, *Proceedings of the 7th IFAC International Symposium on CACSD*, Gent, pp. 341–346, 1997a.

Krohling, R.A., Synthesis of PID controllers for disturbance rejection: A real-coded genetic algorithms approach, *Proceedings of the European Conference on Intelligent Techniques*, Aachen, pp. 862–868, 1997b.

Krohling, R.A., Design of a PID controller for disturbance rejection: A genetic optimization approach, *Proceedings of the 2nd IEE/IEEE International Conference on Genetic Algorithms in Engineering Systems: Innovations and Applications*, Glasgow, pp. 498–503, 1997c.

Krohling, R.A., Genetic algorithms for synthesis of mixed H_2/H_∞ fixed-structure controllers, *Proceedings of the 13th IEEE International Symposium on Intelligent Control*, Gaithersburg, pp. 30–35, 1998.

Krohling, R.A., Jaschek, H., and Rey, J.P., Designing PI/PID controllers for a motion control system based on genetic algorithms, *Proceedings of the 12th IEEE International Symposium on Intelligent Control*, Istanbul, pp. 125–130, 1997a.

Krohling, R.A. and Rey, J.P., Design of optimal disturbance rejection PID controllers using genetic algorithms, *IEEE Trans. on Evolutionary Computation*, 5, 1, pp. 78–82, 2001.

Lo Bianco, C.G. and Piazzi, A., Mixed H_2/H_∞ fixed structure control via semi-infinite optimization, *Proceedings of the 7th IFAC International Symposium on CACSD*, Gent, pp. 329–334, 1997.

Man, K.F., Tang, K.S., and Kwong, S., Genetic algorithms: Concepts and applications, *IEEE Transactions on Industrial Electronics*, 43, 5, 519–533, 1996.

Man, K.F. et al., *Genetic Algorithms for Control and Signal Processing*, Springer-Verlag, Berlin, 1997.

Michalewicz, Z., *Genetic Algorithms + Data Structure = Evolution Programs*, Springer-Verlag, Berlin, 1996.

Onnen, C. et al., Genetic algorithms for optimization in predictive control, *Control Engineering in Practice*, 5, 10, pp. 1363–1372, 1997.

Patton, R.J. and Liu, G.P., Robust control design via Eigenstructure assignment, genetic algorithm and gradient-based optimization, *IEE Proceedings*, Part D Control Theory Application, 141, 3, pp. 202–208, 1994.

Porter II, L.M. and Passino, K., Genetic model reference adaptive control, *Proceedings of the 9th IEEE International Symposium on Intelligent Control*, Columbus, pp. 219–224, 1994.

Schneider, F., Geschlossene Formel zur Berechnung der quadratischen und der zeitbeschwerten quadratischen Regelfläche für kontinuierliche und diskrete Systeme, *Regelungstechnik*, 14, 4, 159–166, 1966.

Wang, P. and Kwok, D.P., Optimal design of PID processes controllers based on genetic algorithms, *Control Engineering in Practice*, 2, 4, 641–648, 1994.

Westcott, J.H., The minimum-moment-of-error-squared criterion: a new performance criterion for Servo mechanisms, *IEE Proceedings*, pp. 471–480, 1954.

Wolfram, S., *Mathematica: A System for Doing Mathematics by Computer*, Addison-Wesley, New York, 1988.

chapter four

Predictive and variable structure control designs

4.1 Model-based predictive controllers

Model-based predictive control is one of the main control strategies employed in applications of advanced control in the process industries. One of the primary reasons for industrial success of model-based control is that hard constraints can be enforced on the process. The technology of model-based predictive control algorithms was developed originally for power systems and petroleum refineries but actually can be met in many successful applications, such as chemical processes, food processing, paper manufacturing, automobile industry, metallurgy, and aerospace applications. In Qin and Badgwell (1997) are presented 2233 commercial applications of predictive controllers.

Predictive controllers present any advantages about other control methods. Model-based predictive control algorithms are versatile and robust in order to deal with plants with complex characteristics. These controllers are attractive, because they allow dealing with challenging control problems such as nonminimum phase behavior, unknown and possibly variable time delay, plant–model mismatch, and open-loop unstable processes. The choice of advanced control algorithms, especially the predictive control algorithms, is based on the consideration that they present adequate performance for nonlinear processes and, also, that they are a reality in controlling industrial processes. Other interesting characteristics of model-based predictive control are the ability to treat knowledge of future requirements for the process state in terms of a predefined tracking–reference signal, and its ability to cope with hard constraints on inputs and states.

Robustness and stability are also highly desirable properties for process control systems. Analysis of robustness and stability of model-based predictive control become more difficult because of its rolling optimization. Qualitatively speaking, a controller is robust if it results in actual closed-loop behavior that does not deviate unacceptably to a nominal process behavior.

For example, a model-based controller results in robust closed-loop stability if the closed-loop is stable, even if there is a discrepancy between the model utilized by the controller and the process behavior. The extent of such discrepancy for the closed-loop stability maintained corresponds to the degree of robustness of that controller.

4.1.1 Basic concepts and algorithms

The design principle behind predictive control is conceptually simple. A multistep (or long-range) cost functional is defined such that it reflects the tracking error between the reference and the plant's output, starting from the current time instant up to a chosen time horizon in the future. Such cost is minimized at each time step relative to the control signal or to some adjustable parameters of a controller. A suitable plant description (model or predictor) is used for forecasting.

There are, in the literature, several examples of model-based predictive control technologies applied in academia and industrial environments, such as MPHC (Model Predictive Heuristic Control), DMC (Dynamic Matrix Control), Predictor-Based Self-Tuning Control, EHAC (Extended Horizon Adaptive Control), MAC (Model Algorithmic Control), MUSMAR (Multistep Multivariable Adaptive Control), EPSAC (Extended Prediction Self-Adaptive Control), and others.

Theory has contributed to the development of model predictive control mainly in its discovery of conditions that ensure closed-loop stability. The control task of the model-based predictive control is to give a series to control signals minimizing a quadratic deviation between a reference signal and the system output in a given prediction horizon. According to the receding horizon strategy, only the first control value is applied, and the procedure is repeated. Three concepts usually found in model predictive control are noted: explicit use of a mathematical model to predict the process output; calculation of a control sequence to optimize a quadratic performance index; and a receding horizon strategy, so that at each instant, the horizon is moved toward the future.

Some of the advantages of the technique are given, and it is noted that it has the disadvantage that an appropriate process model is needed. The success of model-based predictive control is closely related to the quality of the model. That is, unless one is able to find an accurate model of a process, it is unlikely that the associated predictive controller will give adequate performance.

Several optimization and identification techniques can be utilized to obtain an adequate process model for predictive controllers applications, such as successive quadratic programming for constrained nonlinear models (Sandoz et al., 2000), recursive least squares for adaptive predictive controllers (Zhaoli et al., 2000), gradient-based modified Marquardt and finite difference methods for nonlinear models (Pröll and Karim, 1994), neural networks (Saint-Donat et al., 1991), and fuzzy systems (Fischer et al., 1997).

The conventional iterative optimization methods are sensitive to the initialization of the algorithm and usually lead to unacceptable solutions due to convergence to local optima. Genetic algorithms can be applied as an alternative to classical optimization methods for model-based predictive control design. In the next sections, the formulation and an overview of generalized predictive control based on genetic algorithms optimization are presented.

4.1.2 Generalized predictive control

Clarke et al. (1987a, 1987b) presented a predictive control method called the Generalized Predictive Control (GPC), which is a reasonable representative of the model predictive control techniques and has become one of the most popular predictive methods both in industry and academia, being successfully implemented in many industrial applications.

In the GPC algorithm, a control input is generated that minimizes a quadratic cost function consisting of a weighted sum of errors between desired and predicted future system outputs and future predicted control increments. The predictions are obtained from a mathematical model of process dynamics.

4.1.2.1 Formulation and design of GPC

This section presents the synthesis and equations of generalized predictive control of indirect adaptive type. In indirect (or explicit) types of adaptive control, the mathematical model of process is identified (model parameters), and the identification is followed by a calculation of the GPC law parameters. The formulations of GPC design presented in the following are based on Clarke et al. (1987a, 1987b), Clarke (1988), and Bitmead et al. (1990).

A multistep-ahead predictive controller can utilize a mathematical model for process of a CARIMA (Controlled Auto-Regressive Integrated Moving Average) type that is given by the following:

$$A(q^{-1})y(t) = q^{-d}B(q^{-1})u(t) + C(q^{-1})\xi(t) / \Delta \tag{4.1}$$

where $A(q^{-1})$ and $B(q^{-1})$ are polynomials in the delay operator q^{-1}, $y(t)$ and $u(t)$ are output and control variables, respectively, Δ represents the differentiating operator $1 - q^{-1}$, and $\xi(t)$ represents an uncorrelated random sequence. Equation (4.1) can be written in the following form:

$$A(q^{-1})y(t) = q^{-d}B(q^{-1})\Delta u(t) + C(q^{-1})\xi(t) \tag{4.2}$$

where $A(q^{-1}) = 1 + a_1 q^{-1} + ... + a_{na}q^{-na}$, $B(q^{-1}) = b_0 + b_1 q^{-1} + ... + b_{nb}q^{-nb}$, and $C(q^{-1}) = c_0 + c_1 q^{-1} + ... + c_{nc}q^{-nc}$.

Given a process model, the predictive control algorithm projects future outputs given currently available data $\{y(j), u(j-1); j \le t\}$ up to a prediction

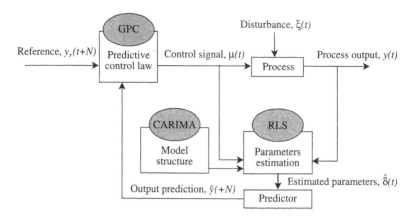

Figure 4.1 Schematic diagram of an adaptive GPC (indirect approach).

horizon, based on a set of assumptions about present and future controls $\{u(j); j \geq t\}$. A recursive least-squares algorithm (RLS) (Ljung, 1987) is utilized to update estimates of the parameters of the CARIMA model, as presented in Figure 4.1.

The control law of GPC is obtained by minimization of a cost function, J, Equation (4.3), subject the constraint of Equation (4.4),

$$J(N_1, N_2, NU, \Gamma) = E\left\{ \sum_{j=N_1}^{N_2} \left[y(t+j) - y_r(t+j) \right]^2 + \Gamma \sum_{j=1}^{N_u} \Delta u^2(t+j-1) \right\} \quad (4.3)$$

$$\Delta(q^{-1})u(t+j-1) = 0, \quad j \geq N_u \quad (4.4)$$

where N_1 is the minimum output horizon, N_2 is the maximum output horizon, N is the control horizon, and Γ is the weighting sequence. The variables $y(t+j)$ and $y_r(t+j)$ represent the output signal and the reference signal, respectively, with j step ahead and $\Delta u(t+j-1)$ as the increment of the control signal in the instant $(t+j-1)$. The prediction horizons and control weighting are main parameters of tuning the GPC.

With Equations (4.1 – 4.3), the following polynomial identity can be obtained

$$C(q^{-1}) = A(q^{-1})\Delta E_j(q^{-1})u(t) + q^{-j}F_j(q^{-1}) \quad (4.5)$$

where the coefficients of polynomials are $E_j(q^{-1}) = 1 + e_1q^{-1} +\ldots+ e_{ne}q^{-ne}$, F_j $(q^{-1}) = f_0 + f_1q^{-1} +\ldots+ f_{nf}q^{-nf}$, $ne = j - 1$, and $nf = \max(na, nc - j)$.

These coefficients are determined by knowledge of prediction interval j and polynomials $A(q^{-1})$ and $C(q^{-1})$. The multiplication of mathematical model equation of process, Equation (4.2), by $q^jE_j(q^{-1})$, gives

$$E_j(q^{-1})A(q^{-1})\Delta y(t+j) = E_j(q^{-1})B(q^{-1})\Delta u(t+j-d) +$$
$$E_j(q^{-1})C(q^{-1})\xi(t+j) \tag{4.6}$$

and, substituting Equation (4.5) by (4.6), the following equation is obtained:

$$C(q^{-1})y(t+j) = E_j(q^{-1})B(q^{-1})\Delta u(t+j-d) +$$
$$F(q^{-1})y(t) + E_j(q-1)C(q^{-1})\xi(t+j) \tag{4.7}$$

Equation (4.7) can be written as follows:

$$y(t+j) = \frac{F_j(q^{-1})}{C(q^{-1})}y(t) + \frac{G'_j(q^{-1})}{C(q^{-1})}\Delta u(t+j-d) + E_j(q-1)\xi(t+j \tag{4.8}$$

where $G'_j(q^{-1}) = E_j(q^{-1})B(q^{-1})$. The noise is uncorrelated of measured signals in time t. Consequently, the predicted output of $y(t+j)$ is as follows:

$$\hat{y}(t+j) = \frac{F_j(q^{-1})}{C(q^{-1})}y(t) + \frac{G'_j(q^{-1})}{C(q^{-1})}\Delta u(t+j-d) \tag{4.9}$$

Using the second polynomial identity,

$$G'_j(q^{-1}) = C(q^{-1})G_j(q^{-1}) + q^{-j}\overline{G}_j(q^{-1} \tag{4.10}$$

and substituting in Equation (4.8) gives

$$\hat{y}(t+j) = \frac{F_j(q^{-1})}{C(q^{-1})}y(t) + \frac{\overline{G}_j(q^{-1})}{C(q^{-1})}\Delta u(t-d) + G_j(q^{-1})\Delta u(t+j-d) \tag{4.11}$$

where the elements g_i of $G_j(q^{-1})$ are composed by an impulsive response of the process model that corresponds to the terms of division $B(q^{-1})/A(q^{-1})\Delta$. Based on Equation (4.11), the prediction of the free response of the system is eliminated, i.e.,

$$y(t+j/t) = \frac{F_j(q^{-1})}{C(q^{-1})}y(t) + \frac{G'_j(q^{-1})}{C(q^{-1})}\Delta u(t-d) \tag{4.12}$$

The vector f is formed by predictions of free response:

$$\mathbf{f} = [\hat{y}(t+N_1/t) \quad \hat{y}(t+N_1+1/t) \quad \cdots \quad \hat{y}(t+N_2/t)\,]^{\mathsf{T}} \qquad (4.13)$$

and the vector of future incremental control

$$\Delta \mathbf{U} = [\Delta u(t) \quad \Delta u(t+1) \quad \cdots \quad \Delta u(t+N_u-1)\,]^{\mathsf{T}} \qquad (4.14)$$

Equation (4.9) can be represented in vector form:

$$\hat{\mathbf{Y}} = \mathbf{G}\Delta \mathbf{U} + \mathbf{f} \qquad (4.15)$$

where

$$\hat{\mathbf{Y}} = [\hat{y}(t+N_1) \quad \hat{y}(t+N_1+1) \quad \cdots \quad \hat{y}(t+N_2)\,]^{\mathsf{T}}$$

$$\mathbf{G} = \begin{bmatrix} g_{N_1-d} & \cdots & g_0 \; 0 \; 0 \cdots & 0 \\ g_{N_1-d+1} & \cdots & g_1 \, g_0 \; 0 \cdots & 0 \\ \vdots & \ddots & \ddots & \vdots \\ \vdots & & \ddots & g_0 \\ \vdots & \cdots & & \vdots \\ g_{N_2-d} \; g_{N_2-d-1} & \cdots & & g_{N_2-N_u-d+1} \end{bmatrix}$$

The matrix *G* has dimension $(N_2-N_1+1) x N_u$. This matrix considers the supposition of $\Delta u(t+j-1) = 0$, $\forall j > N_u$ for penalization of the control ahead of this horizon and reduction of the computational cost of control algorithm. The cost function of GPC can be represented in vector form, i.e.,

$$J(N_1,N_2,NU,\Gamma) = [\mathbf{Y}-\mathbf{Y_r}]^{\mathsf{T}}[\mathbf{Y}-\mathbf{Y_r}] + \Delta \Delta \mathbf{U}^{\mathsf{T}}\Delta \qquad (4.16)$$

where $Y_r = [y_r(t+N_1)\, y_r(t+N_1+1)\ldots y_r(t+N_2)]^{\mathsf{T}}$.

In this way, the minimization of Equation (4.16) is realized, with obtaining the following control law:

$$\Delta \mathbf{U} = [\mathbf{G}^{\mathsf{T}}\mathbf{G}+\Gamma \mathbf{I}]^{-1}\mathbf{G}^{\mathsf{T}}[\mathbf{Y_r}-\mathbf{f}] \qquad (4.17)$$

In practice, only the first control signal is applied, and with each iteration, a new optimization problem is resolved. Consequently, the control law is calculated by the following:

$$u(t) = u(t-1) + \Delta u(t) \qquad (4.18)$$

In order to get the best performance out of the GPC, there needs to be a synergy between the model identification algorithm and the control design so that they are mutually supportive. Here, it is demonstrated how a particular, rather unnoticed approach, can produce surprisingly good results compared to the more usual techniques adopted. The method uses models specifically identified and designed for a given prediction horizon.

Unfortunately, many parameters have to be tuned in order to achieve efficient control law for the GPC. The goal of the next chapter is to concentrate on tuning methodology of GPC in order to derive a nearly automatic design strategy based on genetic algorithms.

GPC has recently received much attention from theoreticians and practitioners in diverse fields. However, the GPC has limitations with complex processes for some of the choices of its design parameters. So far, only a few guidelines related to tuning of the parameters of GPC have been provided in the literature (Rani and Unbehauen, 1996; Liu and Wang, 2000; Al-Ghazzawi et al., 2001). In fact, these parameters are generally determined by the design's experience. From the process control point of view, it is difficult to find out the optimal parameters for the control system based on the quadratic performance index (Liu and Wang, 2000). Consequently, a challenging problem is how to tune the parameters of GPC design.

4.1.2.2 Overview of optimization of GPC design by genetic algorithms

In the last years, much attention has been paid to the model-based predictive control design of genetic algorithms. Genetic algorithms are a versatile and powerful tool for model-based predictive control design, mainly in controlling the processes with relatively slow dynamics and long control horizons. Several approaches have been proposed for optimization by genetic algorithms for configuration predictive control. Examples of GPC design structures relevant, based on genetic algorithms, are the following: determination of the optimum future control sequences; optimization of the tuning parameters, such as control signal weighting, prediction horizons of initial and final output, and control horizon; minimization of a nonlinear cost function; genetic identification of process structure to be controlled; and genetic identification in predictive control design.

Among the tuning parameters approaches, the following must be mentioned (see Figure 4.2): tuning parameters of the prediction component; tuning parameters of the cost function; and optimization of the identification procedure. All of these approaches can be adopted for the closed-loop system to achieve good dynamics and robustness.

Any examples of genetic design of GPC treated in the literature are presented in the following. Goggos and King (1996) present the design of predictive controllers of DMC type based on parameters optimization by genetic algorithms. Martínez et al. (1996) present a comparative study of classical methods versus genetic optimization applied in generalized predictive controller design. Woolley et al. (1997) present the design of an advanced control package based on genetic optimization. Onnen et al.

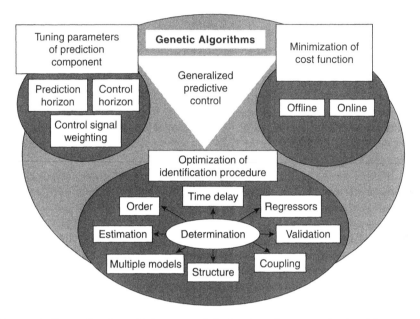

Figure 4.2 Generalized predictive control design based on genetic algorithms.

(1997) investigate control accuracy and computational costs of genetic algorithms and branch-and-bound method for optimization in predictive control. Shin and Park (1998) present a predictive control approach based on genetic optimization and identification based on neural networks. Rauch and Harremoës (1999) present an approach of nonlinear model predictive control based on genetic algorithms for optimization of the urban wastewater performance under dynamic loading from rain. Coelho and Coelho (2000) propose the multivariable predictive control design based on neural network identification and optimization based on simplex and evolutionary hybrid method (memetic algorithm). Mahfouf et al. (2000) propose a multiobjective genetic algorithm for optimizing the tuning parameters related to the GPC design with performance indices based on fuzzy system.

4.2   Variable structure control systems

4.2.1   Introduction

The classical and modern control theories for linear processes are considered a well-defined scientific discipline with powerful techniques for analyzing and designing controllers. However, in many engineering applications, the application of linear control theory can present problems caused by the fact that: (i) a linear mathematical model of the process is needed, and finding one is not a trivial problem in many cases; (ii) modeling difficulties have forced control engineers to use simplified or linearized models, which are often inaccurate and vulnerable to parameter inaccuracy (or structured uncertainty) and unmodeled dynamics (or unstructured uncertainty), resulting in degraded controlled system performance (Li et al., 1996); (iii) because of changing environmental conditions, such as ambient humidity and temperature, most processes are time varying; and (iv) most industrial processes present complexities, mainly chemical processes, such as inherent process nonlinearity, process model uncertainty, process constraints and dead time, unmeasurable process variables caused by limits in actuators, sensor ranges, and other elements.

Currently, several approaches have been adopted in control engineering to treatment of modeling and identification difficulties, such as predictive control (Chisci et al., 2001), adaptive control (Li and Wozny, 2001), robust control (Gustafsson and Mäkilä, 2001), neural networks (Zhao et al., 2001), fuzzy systems (Wong et al., 2000), evolutionary computation (da Silva et al., 2000), and hybrid intelligent systems (Coelho and Coelho, 1998).

A relevant methodology for combating modeling and identification difficulties involves robust control, such as sliding mode control system that is also known as variable structure system control. Sliding mode control systems are characterized by a discontinuous feedback control law that switches the structure of the system during the evolution of the state vector in order to maintain the state trajectories in a predefined subspace (Li et al., 1996; Utkin, 1977).

Sliding mode control has been promoted as a very robust control algorithm, able to provide robustness to bounded model uncertainty and fast response and accuracy in the presence of widely variable operating conditions or nonlinear systems. It is an efficient method of controlling complex nonlinear dynamic systems. Its well-known order reduction property and its insensitivity to disturbances and plant parameter variations makes it an interesting tool for nonlinear robust control. The sliding-mode control has been successfully applied to robot control, aircraft control, load frequency control of power systems, underwater vehicles, automotive transmissions and engines, and others (Hung et al., 1993).

A variable structure system is a dynamical system with a structure that changes in accordance with the current value of its state. A variable structure system can be viewed as a system composed of independent structures together with a switching logic to switch between the structures. The procedure for the design of a sliding mode control can be separated into two parts: design a switching surface to represent the desired system dynamics and design a variable structure control law. By introducing sliding modes into the system, it is possible to stabilize the system while achieving insensitivity to external disturbances and parametric deviations.

There are two phases in a traditional variable structure control system: a reaching phase and a sliding phase. The reaching phase is also called the nonsliding phase, in which the trajectory approaches the sliding surface from an arbitrary initial position within a finite time. The sliding phase ensures that the trajectory asymptotically moves toward the equilibrium point of the switching plane. When the system states are on the sliding phase, the system response will only depend on the predesigned sliding surface parameters and is independent of system dynamics. Hence, the sliding mode control ensures the robustness of the system. Because it is impossible to obtain an accurate model for practical systems, sliding mode control has been widely the effect of model uncertainty and external disturbance.

Originally, the sliding mode control theory was developed in a continuous time domain. Most of the initial efforts of variable structure control research have concentrated on the continuous time case (Emelyanov, 1967; Itkis, 1976; Utkin, 1977). Generally, the continuous time sliding mode approach provides robustness to matched disturbances and system uncertainties.

Due to the fast development of electronics digital and personal computers, it is quite natural to extend the technique of continuous sliding mode control to discrete-time control systems. When the digital controller realizes the control law derived from the traditional continuous-time sliding mode control theory, it will result in some problems. Because the control signal is constant for every sampling period and the sampling rate cannot be infinite, the ideal sliding mode defined in the continuous-time system will not occur in the discrete-time system, even without external disturbances and parameter uncertainties. This can yield chattering or even lead to system instability. In the discrete-time system, the "sliding" phenomenon of the system state at the sliding surface does not exist. Hence, the sliding surface is thought of as a switching surface (Chang and Zhu, 2001).

The implementation of the controller on a digital computer presents difficulties due to limited sampling rate, sample/holds effects, and discretization errors, and it requires a certain sampling interval, and the assumption of an infinite switching time does not hold anymore. For the discrete-time model, the control guaranteeing the stability of the sliding mode must be upper and lower bounded. These bounds are shown to depend on the sampling period and on the uncertainties in the system (Kotta, 1989).

Discrete sliding mode control (quasi-sliding mode control) has received attention recently (Furuta et al., 1989; Hung et al., 1993; Bartolini et al., 1995; Koshkouei and Zinober, 1996; Haskara et al., 1997; Bartoszewicz, 1998; Lee et al., 1999). These controllers are formulated numerous ways for systems with different kinds of uncertainties using switching or nonswitching types of techniques. The existence of a sliding mode cannot be guaranteed in the presence of uncertainties in the discrete-time systems. The characteristics of discrete-time variable structure control systems differ from those of continuous-time variable structure systems in two aspects. First, the discrete-time variable structure systems can only undergo quasi-sliding modes. Second, when the state does reach the switching surface, the subsequent discrete-time switching cannot generate the equivalent control to keep the state on the surface (Hung et al., 1993).

To design a variable structure system controller for discrete-time system (quasi-sliding mode controller), instead of sliding mode, a sliding sector was proposed in Furuta (1990), the discrete-time variable structure system controller was designed to move system state from the outside to the inside of the sliding sector, where the system is stable (Furuta and Pan, 1996).

Robust control (such as the quasi-sliding mode control) and adaptive control have been viewed as two methodologies competing for controller design in the presence of structured and unstructured uncertainties. Recent developments indicate that these two techniques complete each other, and the key for understanding their powerful characteristics has been provided by progress made in several adaptive robust control approaches for application of processes with deadtime (Oucheriah, 2001), deadzone (Xu and Viswanathan, 2000), unknown plant orders (Veres and Sokolov, 1998), and unknown parameters (Yoneyama et al., 1997).

This chapter presents discrete-type variable structure control (quasi-sliding mode control) design proposed by Furuta et al. (1989) and its application to adaptive control with estimated parameters by least mean squares algorithm. In the proposed approach, the quasi-sliding mode control system uses genetic algorithms for optimization of design parameters of control law. The genetic algorithm for design configuration of a discrete sliding mode control system provides an automatic and optimized selection of sliding slopes and controller parameters, offering superior performances in transient, steady-state, and robustness to manual designs.

Section 4.2.2 gives the basic concepts and design of quasi-sliding mode control, respectively. The optimization of controller parameters based on genetic algorithms is presented in Section 4.2.3.

4.2.2   Basic concepts and controller design

As an alternative design algorithm of the continuous sliding mode control, a sliding sector has been proposed to replace the sliding mode in the design of variable structure control for a chattering free controller and for the implementation in discrete-time control systems (Furuta, 1990).

The control law of quasi-sliding mode control is known to consist of the following steps: determine the switching function, s, such that the sliding mode on switching plane is stable and determine the control law such that a reaching condition is satisfied. As a result, the closed-loop is globally stable. Gao et al. (1995) consider that a discrete variable system control should have three attributes: (1) starting from any initial state, the trajectory will move monotonically toward the switching plane and cross it in finite time; (2) once the trajectory has crossed the switching plane the first time, it will cross the plane again in every successive sampling period, resulting in a zigzag motion about the switching plane due to limited switching frequency; and (3) the size of each successive zigzagging step is nonincreasing, and the trajectory stays within a specified band.

Because the sliding mode features of discrete variable system control are different from those of continuous variable structure control approaches, the sliding mode for discrete variable system control should be formally defined. The following three definitions are given to accomplish this (Gao et al., 1995):

Definition 1: The motion of a discrete variable structure control system satisfying attributes (1) and (3) is called *quasi-sliding mode control*. The specified band that contains the quasi-sliding mode control band is defined by

$$\{|x| - \Delta < s(x) < \Delta\} \tag{4.19}$$

where 2Δ is the width of the band. Bartoszewicz (1998) describes the quasi-sliding mode as a motion of the system, such that its state always remains in a certain band around the sliding hyperplane.

Definition 2: The quasi-sliding mode becomes an *ideal quasi-sliding mode* when $\Delta = 0$.

Definition 3: A discrete variable structure control system is said to satisfy a *reaching condition* if the resulting system possesses all three attributes: (1), (2), and (3).

The self-tuning control based on variable structure systems design proposed by Furuta et al. (1989) is utilized in the simulations of the next chapter. In this case, the genetic algorithms are utilized in the variable structure parameters design. This design is utilized in second-order processes with disturbances and unmodeled dynamics. The mathematical model of process is represented by equations of ARMA (Auto-Regressive Moving Average) type:

$$A(q^{-1})y(t) = q^{-d}B(q^{-1})u(t) + \xi(t) \tag{4.20}$$

with

$$A(q^{-1}) = 1 + a_1 q^{-1} + \dots + a_{na} q^{-na}$$

$$B(q^{-1}) = b_0 + b_1 q^{-1} + \dots + b_{nb} q^{-nb}$$

where $A(q^{-1})$ and $B(q^{-1})$ are polynomials in the delay operator q^{-1}, d is time delay, $y(t)$ and $u(t)$ are output and control variables, respectively, and $\xi(t)$ represent an uncorrelated random sequence. A recursive least-squares algorithm (Ljung, 1987) is utilized to update estimates of parameters $\hat{\theta} = \{\hat{a}_1, \hat{a}_2, \hat{b}_0, \hat{b}_1\}$ of the ARMA model of second order.

The definition of the sliding hypersurfaces is given by the following:

$$s(t+1) = e(t+1) + k_1 e(t) + k_2 e(t-1) = 0 \tag{4.21}$$

and the error is given by the following equation:

$$e(t) = y(t) - y_r(t) \tag{4.22}$$

where k_1 and k_2 are determined so that the error $e(t)$ is stable on the hyper-surface $s(t) = 0$. The control signal is chosen to be in the form of

$$\begin{bmatrix} e(t) \\ e(t+1) \end{bmatrix} = \begin{bmatrix} 0 & 1 \\ -k_2 & -k_1 \end{bmatrix} \begin{bmatrix} e(t-1) \\ e(t) \end{bmatrix} + \begin{bmatrix} 0 \\ 1 \end{bmatrix} v'(t) \tag{4.23}$$

$$v'(t) = e(t) + k_1 e(t-1) + k_2 e(t-2) + \begin{bmatrix} f_1 & f_2 \end{bmatrix} \begin{bmatrix} e(t-1) \\ e(t) \end{bmatrix} \tag{4.24}$$

where the last component $\begin{bmatrix} f_1 & f_2 \end{bmatrix} \begin{bmatrix} e(t-1) \\ e(t) \end{bmatrix}$ is the switching term of the equa-

tion, and the component $e(t) + k_1 e(t-1) + k_2 e(t-2)$ driving the state along the sliding hypersurfaces. A relevant definition is a positive definite function $V(t)$ given by

$$V(t) = \frac{1}{2} s(t)^2 \tag{4.25}$$

and the definition of $\Delta s(t + 1)$ as the difference

$$\Delta s(t+1) = s(t+1) - s(t) \tag{4.26}$$

From Equation (4.26) is obtained the following equation:

$$V(t+1) = V(t) + 2s(t)\Delta s(t+1) + \Delta s(t+1)^2 \tag{4.27}$$

The control objective will be to make $V(t)$ decrease along the switching hypersurfaces. From Equation (4.27), the following condition is obtained:

$$s(t)\Delta s(t+1) < -\frac{1}{2} [\Delta s(t+1)]^2 \tag{4.28}$$

Using Equation (4.26), the control signal has the following equation:

$$\Delta s(t+1) = \begin{bmatrix} f_1 & f_2 \end{bmatrix} \begin{bmatrix} e(t-1) \\ e(t) \end{bmatrix} \tag{4.29}$$

$$\Delta s(t+1)s(t) = f_1 e(t-1)s(t) + f_2 e(t)s(t)$$

$$< -\delta_1 f_0 - \delta_2 f_0 \tag{4.30}$$

$$< -\frac{1}{2}\left[f_0^2 e(t-1)^2 + 2f_0^2 e(t-1)e(t) + f_0^2 e(t)^2\right]$$

where

$$\delta_i = \frac{1}{2}\left[f_0 e(t+i-2) + f_0|e(t-1)\|e(t)|\right], \quad f_0 > 0, \ i = 1,2 \tag{4.31}$$

$$f_i = \begin{cases} f_0 & \text{if} \quad e(t+i-2)s(t) < -\delta_i \\ 0 & \text{if} \quad e(t+i-2)s(t) < \delta_i \quad, \\ -f_0 & \text{if} \quad e(t+i-2)s(t) > \delta_i \end{cases} \qquad f_0 > 0, \ i = 1,2 \tag{4.32}$$

A cost function J that will optimized is

$$J = p\left[y(t+1) - v(t) - v'(t)\right]^2 + r\left[u(t) - u(t-1)\right]^2 \tag{4.33}$$

The following control law makes the cost function (4.33) minimal. Using the estimated parameters $\hat{\theta}$, the above control law becomes:

$$u(t) = \frac{1}{\hat{b}_1 + r}\{y_r(t+1) - \hat{a}_1 y(t) - \hat{a}_2 y(t-1) - \left[\hat{b}_2 - r\right]u(t-1) - \left[1 - k_1\right]e(t)$$

$$- \left[k_1 - k_2\right]e(t-1) - k_2 e(t-2) - \left[f_1 \ f_2\right]\left[\begin{array}{c} e(t) \\ e(t-1) \end{array}\right]\} \tag{4.34}$$

4.2.3   Overview of optimization of variable structure control design by genetic algorithms

In recent years, much attention has been paid to the optimization of control parameters of variable structure control systems based on genetic algorithms. Any examples are presented in the following. Li et al. (1996) develop the design of sliding mode control systems based on genetic algorithms. Jeon et al. (1998) apply genetic algorithms to optimize the control gains of a three-loop controller for a pneumatic servo cylinder drive. Al-Duwaish and Al-Hamouz (1998) treat the selection of parameters of a sliding mode controller based on genetic algorithms. Kaynak and Rudas (1998) present a tutorial about combinations of computational intelligence methodologies, such as fuzzy systems, neural networks, and genetic algorithms in sliding mode control design. Cheong et al. (1999) present experimental results of sliding

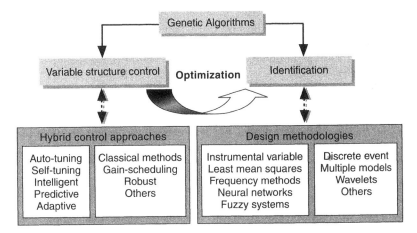

Figure 4.3 Variable structure control designs based on genetic algorithms.

mode control based on genetic algorithms for micro-drilling productivity enhancement. Tamal et al. (1999) propose a neural-fuzzy approach for sliding mode control. Go and Lee (2000) relate a self-tuning fuzzy inference method by the genetic algorithm in the fuzzy-sliding mode control for a polishing robot. McGookin et al. (2000) present the performance of sliding mode controller with optimization using genetic algorithms for regulating the motion of a ship model. Su et al. (2001) apply a stable adaptive fuzzy sliding mode continuous control with genetic optimization for a class of nonlinear processes. The research of control design of variable structure systems can be resumed in Figure 4.3.

To implement discrete-time variable structure control based on genetic algorithms optimization, three important problems are considered. They are choice of dynamic sliding surface for the process, computation of the discrete-time dynamic sliding surface variable, and self-tuning of the switching control magnitude to reduce chattering. In the next chapter, a design methodology of quasi-sliding mode control based on genetic algorithms is presented.

References

Al-Duwaish, H.N. and Al-Hamouz, Z.M., A genetic approach to the selection of the variable structure controller feedback gains, *Proceedings of the IEEE International Conference on Control Applications*, Trieste, Italy, vol. 1, pp. 227–231, 1998.

Al-Ghazzawi, E. Ali, Nouth, A., and Zafiriou, E., Online tuning strategy for model predictive control, *Journal of Process Control*, 11, 265–284, 2001.

Bartolini, G., Ferrara, A., and Utkin, V.I., Adaptive sliding mode control in discrete-time systems, *Automatica*, 31, 5, 769–773, 1995.

Bartoszewicz, A., Discrete-time quasi-sliding mode control strategies, *IEEE Transactions on Industrial Electronics*, 45, 4, 633–637, 1998.

Bitmead, R.R., Gevers, M., and Wert, V., *Adaptive optimal control — the thinking man's GPC*, Prentice-Hall, Englewood Cliffs, New Jersey, 1990.

Chang, K.-M. and Zhu, Z.-H., Discrete-time sliding mode controller design with weak pseudo sliding condition, *Journal of Mathematical Analysis and Applications*, 258, 536–555, 2001.

Cheong, M.S., Cho, D.-W., and Ehmann, K.F., Identification and control for microdrilling productivity enhancement, *International Journal of Machine Tools & Manufacture*, 39, 1539–1561, 1999.

Chisci, L., Rossiter, J.A., and Zappa, G., Systems with persistent disturbances: Predictive control with restricted constraints, *Automatica*, 37, 7, 1019–1028, 2001.

Clarke, D.W., Application of generalized predictive control to industrial processes, *IEEE Control Systems*, April, 49–55, 1988.

Clarke, D.W., Mohtadi, C., and Tuffs, P.S., Generalized predictive control — part I: the basic algorithm, *Automatica*, 23, 2, 137–148, 1987a.

Clarke, D.W., Mohtadi, C., and Tuffs, P.S., Generalized predictive control — part II: extensions and interpretations, *Automatica*, 23, 2, 149–160, 1987b.

Coelho, L.S. and Coelho, A.A.R., Computational intelligence in process control: Fuzzy, evolutionary, neural, and hybrid approaches, *International Journal of Knowledge-Based Intelligent Engineering Systems*, 2, 2, 80–94, 1998.

Coelho, L.S. and Coelho, A.A.R., Multivariable predictive control based on neural network model and simplex-evolutionary hybrid optimization, *Soft Computing in Industrial Applications*, Suzuki, Y. et al., Eds., Springer, London, UK, 2000.

Da Silva, W.G., Acarnley, P.P., and Finch, J.W., Application of genetic algorithms to the online tuning of electric drive speed controllers, *IEEE Transactions on Industrial Electronics*, 47, 1, 217–219, 2000.

Emelyanov, S.V., *Variable Structure Systems*, Moscow: Nauka (in Russian), 1967.

Fischer, M., Schmidt, M., and Kavsek-Biasizzo, K., Nonlinear predictive control based on the extraction of step response models from Takagi-Sugeno fuzzy systems, *American Control Conference*, Albuquerque, New Mexico, 1997.

Furuta, K., Sliding mode control of a discrete system, *System & Control Letters*, 14, 145–152, 1990.

Furuta, K. and Pan, Y., Design of discrete-time VSS controller based on sliding sector, *Proceedings of IFAC 13th Triennial World Congress*, San Francisco, California, 487–492, 1996.

Furuta, K., Kosuge, K., and Kobayashi, K., VSS-type self-tuning control of direct-drive motor, *Proceedings of IECON*, Philadelphia, Pennsylvania, 281–286, 1989.

Gao, W., Wang, Y., and Homaifa, A., Discrete-time variable structure control systems, *IEEE Transactions on Industrial Electronics*, 42, 2, 117–122, 1995.

Go, S.J. and Lee, M.C., Fuzzy-sliding mode control with the self tuning fuzzy inference based on genetic algorithm, *Proceedings of the IEEE International Conference on Robotics and Automation*, San Francisco, California, 2124–2129, 2000.

Goggos, V. and King, R.E., Evolutionary predictive control (EPC), *Computers Chem. Engineering*, 20, Suppl., S817–S822, 1996.

Gustafsson, T.K. and Mäkilä, P.M., Modelling of uncertain systems with application to robust process control, *Journal of Process Control*, 11, 3, 251–264, 2001.

Haskara, I., Özgüner, Ü., and Utkin, V., Variable structure control for uncertain sampled data systems, *Proceedings of the 36th Conference on Decision & Control*, San Diego, California, 3226–3231, 1997.

Hung, J.Y., Gao, W., and Hung, J.C., Variable structure control: A survey, *IEEE Transactions on Industrial Electronics*, 40, 1, 2–22, 1993.

Itkis, Y., *Control Systems of Variable Structure*, New York: Wiley, 1976.

Jeon, Y.-S., Lee, C.-O., and Hong, Y.-S., Optimization of the control parameters of a pneumatic servo cylinder drive using genetic algorithms, *Control Engineering Practice*, 6, 847–853, 1998.

Kaynak, O. and Rudas, I.J., The fusion of computational intelligence methodologies in sliding mode control, *Proceedings of the 24th Annual Conference of the IEEE, IECON*, Aachen, Germany, 1, T25–T34, 1998.

Koshkouei, A.J. and Zinober, A.S.I., Discrete-time sliding-mode control design, *IFAC 13th Triennial World Congress*, San Francisco, California, 481–486, 1996.

Kotta, U., Stability of discrete-time sliding mode control systems (comments), *IEEE Transactions on Automatic Control*, 34, 9, 1021–1022, 1989.

Lee, P.-M. et al., Discrete-time quasi-sliding mode control of an autonomous underwater vehicle, *IEEE Journal of Oceanic Engineering*, 24, 3, 388–395, 1999.

Li, P. and Wozny, G., Tracking the predefined optimal policies for multiple-fraction batch distillation by using adaptive control, *Computers & Chemical Engineering*, 25, 1, 97–107, 2001.

Li, Y. et al., Genetic algorithm automated approach to the design of sliding mode control systems, *International Journal of Control*, 63, 4, 721–739, 1996.

Liu, W. and Wang, G., Auto-tuning procedure for model-based predictive controller, *IEEE International Conference on Systems, Man, and Cybernetics*, 5, PAIS, 3421–3426, 2000.

Ljung, L., *System Identification: Theory for the User*, Prentice-Hall, Englewood Cliffs, New Jersey, 1987.

Mahfouf, M., Linkens, D.A., and Abbod, M.F., Multi-objective genetic optimisation of GPC and SOFLC tuning parameters using a fuzzy–based ranking method, *IEE Proc.-Control Theory Appl.*, 147, 3, 344–354, 2000.

Martínez, M., Senté, J., and Blasco, X., A comparative study of classical versus genetic algorithm optimization applied in GPC controller, *IFAC 13th Triennial World Congress*, San Francisco, California, 327–332, 1996.

McGookin, E.W. et al., Ship steering control system optimisation using genetic algorithms, *Control Engineering Practice*, 8, 429–443, 2000.

Onnen, C. et al., Genetic algorithms for optimization in predictive control, *Control Engineering Practice*, 5, 10, 1363–1372, 1997.

Oucheriah, S., Adaptive robust control of a class of dynamic delay systems with unknown uncertainty bounds, *International Journal of Adaptive Control and Signal Processing*, 15, 1, 53–63, 2001.

Pröll, T. and Karim, M.N., Model-predictive pH control using real-time NARX approach, *AIChE Journal*, 40, 2, 269–282, 1994.

Qin, S.J. and Badgwell, T.J., An overview of model predictive control technology, *5th International Conference on Chemical Process Control*, AIChE and CACHE, Kantor, J.C., Garcia, C.E., and Carnahan, B., Eds., Tahoe, California, AIChE Symposium Series 316, 93, 232–256, 1997.

Rani, K.Y. and Unbehauen, H., Tuning and auto-tuning in predictive control, *IFAC 13th Triennial World Congress*, San Francisco, California, 109–114, 1996.

Rauch, W. and Harremoës, P., Genetic algorithms in real time control applied to minimize transient pollution from urban wastewater systems, *Water Res.*, 33, 5, 1265–1277, 1999.

Saint-Donat, J., Bhat, N., and McAvoy, T.J., Neural net based model predictive control, *International Journal of Control*, 54, 6, 1453–1468, 1991.

Sandoz, D.J. et al., Algorithms for industrial model predictive control, *Computing & Control Engineering Journal*, June, 125–134, 2000.

Shin, S.C. and Park, S.B., GA-based predictive control for nonlinear processes, *Electronics Letters*, 34, 20, 1980–1981, 1998.

Su, J.-P., Chen, T.-M., and Wang, C.-C., Adaptive fuzzy sliding mode control with GA-based reaching laws, *Fuzzy Sets and Systems*, 120, 145–158, 2001.

Tamal, Y., Akhmetov, D., and Dote, Y., Novel fuzzy-neural network with general parameter learning applied to sliding mode control systems, *Proceedings of IEEE International Conference on Systems, Man, and Cybernetics*, 1, 376–379, 1999.

Utkin, V.I., Variable structure systems with sliding modes, *IEEE Transactions on Automatic Control*, 22, 2, 212–222, 1977.

Veres, S.M. and Sokolov, V.F., Adaptive robust control under unknown plant orders, *Automatica*, 34, 6, 723–730, 1998.

Wong, C.H. et al., Adaptive fuzzy relational predictive control, *Fuzzy Sets and Systems*, 115, 2, 247–260, 2000.

Woolley, I. et al., Genetic algorithm plug-in controller for the Connoisseur™ control package, *Proceedings of Second IEE/IEEE International Conference on Genetic Algorithms Applications*, GALESIA, Glasgow, UK, 382–387, 1997.

Xu, J.-X. and Viswanathan, B., Adaptive robust iterative learning control with dead zone scheme, *Automatica*, 36, 1, 91–99, 2000.

Yoneyama, J., Speyer, J.L., and Dillon, C.H., Robust adaptive control for linear systems with unknown parameters, *Automatica*, 33, 10, 1909–1916, 1997.

Zhao, H. et al., A nonlinear industrial model predictive controller using integrated PLS and neural net state-space model, *Control Engineering Practice*, 9, 2, 125–133, 2001.

Zhaoli, M. et al., Adaptive predictive control for bilinear systems with model indeterminacy, *Proceedings of the 3rd World Congress on Intelligent Control and Automation*, Hefei, P.R. China, 3205–3207, 2000.

chapter five

Design methods, simulation results, and conclusion

5.1 Optimization of generalized predictive control design by genetic algorithms

GPC is emerging as one of the most effective control techniques in process industries. This is due to the fact that many attributes fundamental in any practical industrial control design may be incorporated in GPC and others' model-based predictive control. However, for an appropriate GPC framework, design is necessary to explore the effect of tuning different design variables.

The inadequate tuning of design variables can result in poor performance, slow response, oscillatory process response, or sometimes an unstable closed-loop system. Several tuning guidelines of design variables for generalized predictive control proposed in the literature have been converted into suitable tuning rules. To obtain improved GPC designs and maintain the benefits of adaptive control techniques, it is necessary to adopt automatic methodologies for adjustment of design parameters. In the next sections, the automatic tuning of parameters of GPC design realized by genetic algorithms is presented.

5.1.1 Design method

The study of this section is concentrated on GPC tuning methodology based on optimization by genetic algorithms. The proposed design methodology provides a systematic way of selecting design parameters compared to trial-and-error methods reported in the literature. The robustness of genetic algorithms for applications of GPC design is due to their capacity to locate the global optimum in a multimodal landscape.

The fundamentals of GPC and the design variables were described in Section 4.1.2.1. Two design conceptions with parameters optimization by genetic algorithms are adopted: adaptive GPC design without constraints and adaptive GPC design with constraints for the control signal.

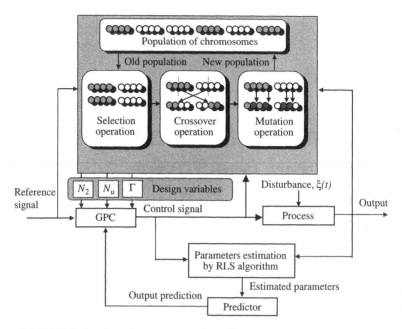

Figure 5.1 GPC design based on genetic algorithms.

5.1.2 Design example

The simulations of GPC design are realized for analysis of servo behavior with a set point change. Genetic algorithms in proposed GPC design optimize any design parameters. The optimized design parameters are the maximum output horizon, N_2, the control horizon, N_u, and the weighted sequence on the control increments, Γ (Figure 5.1). The main parts of the GPC are the prediction model, the reference trajectory, the objective function, and the optimization procedure. The GPC parts and implementation details of a canonical genetic algorithm for a design example are described as follows:

- *Mathematical model*: The discrete mathematical model of the process to be controlled is represented by a CARIMA model. The changes in the polynomials $A(q^{-1})$ and $B(q^{-1})$ are given by the following conditions:
  - Samples 1 to 50:

$$A(q^{-1}) = 1 - 1.807q^{-1} + q^{-2} \text{ and } B(q^{-1}) = 0.0197q^{-1} + 0.0197q^{-2}$$

  - Samples 51 to 100:

$$A(q^{-1}) = 1 - 1.950q^{-1} + 2q^{-2} \text{ and } B(q^{-1}) = 0.9q^{-1} - 0.5q^{-2}$$

where the time delay, d, is considered known and equals zero. The noise variance (an uncorrelated random sequence) is set to 0.001 for all simulations.

- *Prediction model*: The prediction model uses $na = nb = 2$. The parameters of polynomials $A(q^{-1})$ and $B(q^{-1})$ are estimated using a recursive least-squares (RLS) algorithm. An exponential forgetting factor is adopted with values ranging from 0.95 to unity. The estimated parameters are initialized in simulation with $\theta = \{a_1,a_2,b_1,b_2\} = \{0.2,0.2,0.2,0.2\}$. The diagonal of the initial covariance matrix is set to 1000. This conception of design is denominated of indirect adaptive control.
- *Reference trajectory*: The desired reference signal is $y_r(t) = 5$ for the samples 1 to 50, and it is $y_r(t) = 2$ for the samples 51 to 100.
- *Objective function*: The objective function of GPC is given by Equation (4.3), and the control law of Equation (4.17) is obtained by minimization of Equation (4.16).
- *Optimization procedure*: A genetic algorithm is a global search technique that emulates operators of natural selection and natural genetics. A genetic algorithm applies operators inspired by the mechanics of natural selection to a population of binary strings encoding the parameter space. The genetic algorithm consists of three fundamental operators: selection, crossover, and mutation. In the sequel, it illustrates the main features of the genetic algorithm employed in simulations.

- *Fitness function*: The fitness function to be maximized is represented by the following equation:

$$fitness = \frac{k_s}{1 + J_{ga}}$$

where the scale coefficient $k_5 = 300$. The performance index is given by:

$$J_{ga} = k\left[yr(t) - y(t)\right]^2 + 0.1\left[u(t) - u(t-1)\right]^2$$

where $k = \begin{cases} t \text{ (sample 1 to 50)} \\ t - 50 \text{ (samples 51 to 100)} \end{cases}$

- *Chromosome structure and coding*: A canonical genetic algorithm is used in this design example. The population is the set of possible solutions (optimization variables), and each individual of this population is characterized by chromosome-like structures coded in binary strings. The optimization variables N_2 and N_u are coded by using integer variables, and Γ is coded as a real variable. The adopted population size is set to 30 chromosomes, and the search range for each variable is given by:

$$\begin{cases} 2 < N_2 < 8 \\ 1 < N_u < 8 \\ 0 < \Gamma < 10 \end{cases}$$

- *Selection operator*: A set of individuals from the previous population must be selected for reproduction. This selection depends on their fitness values. Individuals with better fitness values will more probably survive. The selection operator used in this example is the roulette wheel approach.
- *Crossover operator*: The crossover operator is applied after the selection operator. Through the crossover operator, genetic material is exchanged between two random strings. The crossover operator for this example employs a single split (simple crossover). This operator is applied in the evolutionary cycle with crossover probability, $p_c = 0.80$.
- *Mutation operator*: The mutation operator is a method for introducing new structures into the population. When mutation occurs, the string position is changed to a different allele selected from the set to possible digits. In the canonical case, zero or one. A prespecified mutation probability is $p_m = 0.10$.
- *Termination criterion*: The adopted termination criterion is the number of generations set to 30. The number of experiments for each GPC design is 25.

5.1.3   Simulation results

5.1.3.1   Case study 1: Adaptive GPC design without constraints
This section shows the results obtained with an adaptive GPC design without constraints. The best result of experiments with genetic algorithms for adaptive GPC design was $N_2 = 8$, $N_u = 8$, and $\Gamma = 1.423$. The fitness of this result was 0.8657. The performance of control genetic design is presented in Figure 5.2.

The GPC design was adequate for the treatment of set point changes of process. However, the performance obtained shows an increase of the control signal variance when set point changes occur. The fitness evolution versus generations and the convergence of three design parameters (N_2, N_u, and Γ) are presented in Figures 5.3 and 5.4, respectively.

5.1.3.2   Case study 2: Adaptive GPC design with constraints for the control signal
This section shows the obtained results with an adaptive GPC design with constraints of control signal. The best results with genetic algorithms for

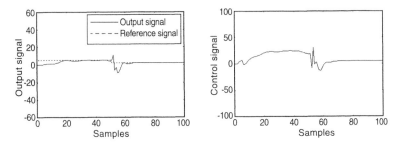

Figure 5.2 Best result of GPC design without constraints based on genetic algorithms.

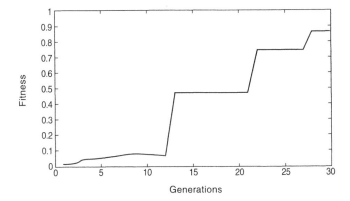

Figure 5.3 Fitness convergence of genetic algorithm for GPC tuning without constraints (best result).

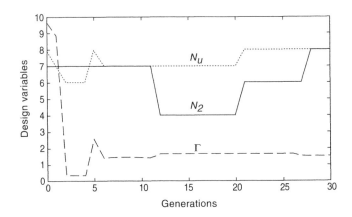

Figure 5.4 Convergence of genetic algorithm for GPC tuning parameters without constraints (best result).

constrained adaptive GPC design are presented in Table 5.1, while the performance of GPC designs with constraints are presented in Figures 5.5 to 5.8.

Table 5.1 Simulation Results for Self-Tuning Quasi-Sliding Mode Control with Constraints

	Design Parameters				Indices		
Constraints	N_2	N_u	Γ	Fitness	N $\sum e(t)^2$ $t = 1$	N $\sum \Delta u(t)^2$ $t = 1$	N $\sum u(t)^2$ $t = 1$
$-20 \leq u(t) \leq 20$	6	5	0.417	1.1431	2614.612	19,638.953	833.057
$-25 \leq u(t) \leq 25$	6	4	0.004	1.422	2099.700	24,736.403	190.794
$-30 \leq u(t) \leq 30$	3	3	0.353	13.304	331.532	23,299.201	215.505
$-35 \leq u(t) \leq 35$	6	5	0.140	0.828	3611.406	28,487.280	651.193

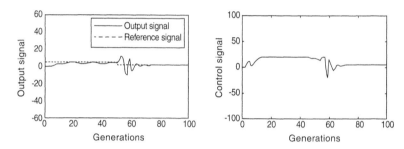

Figure 5.5 Results of GPC design based on genetic algorithms with constraints of control signal $[-20 \leq u(t) \leq 20]$.

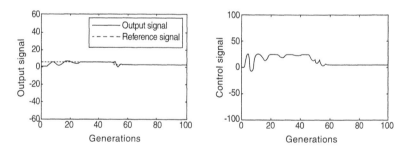

Figure 5.6 Results of GPC design based on genetic algorithms with constraints of control signal $[-25 \leq u(t) \leq 25]$.

The GPC design for $-20 \leq u(t) \leq 20$ presents poor transient behavior for samples 51 to 70. Furthermore, the servo behavior for this controller presents steady state error and oscillations for the tracking of reference $y_r(t) = 5$. The GPC designs with constraints of $-25 \leq u(t) \leq 25$ and $-35 \leq u(t) \leq 35$ also present oscillatory behavior. However, all studied GPC designs with constraints obtain good features of performance for tracking the reference $y_r(t) = 2$.

The proposed optimization design based on genetic algorithms provides an adequate choosing of design parameters of GPC. However, the

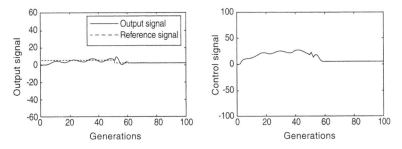

Figure 5.7 Results of GPC design based on genetic algorithms with constraints of control signal $[-30 \le u(t) \le 30]$.

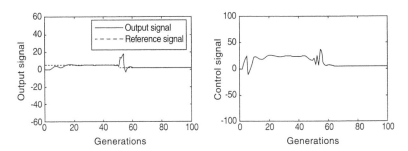

Figure 5.8 Results of GPC design based on genetic algorithms with constraints of control signal $[-35 \le u(t) \le 35]$.

closed-loop performance of the process also depends on design parameters for the recursive identification scheme (initial settings of the covariance matrix and the forgetting factor). In future works, we will investigate the possibility of exploiting various performance indices in multiobjective optimization approaches for achieving performance objectives, such as a fast response of process without oscillation, small rising, and control action with small variance.

5.2 Optimization of quasi-sliding mode control design by genetic algorithms

5.2.1 Design method

The design parameters of variable structure systems are usually determined by trial-and-error approach. This heuristic approach is numerically intensive, especially for a large number of parameters. For an appropriate design of quasi-sliding, mode control law is needed so that the choices of design parameters satisfy certain system performance requirements. The study of this section treats the tuning methodology of self-tuning quasi-sliding mode control design parameters based on genetic optimization.

5.2.2 Design example

The used quasi-sliding mode control design was described in Section 4.2.2. In this design example, two design conceptions with parameters optimization by genetic algorithms are examined: self-tuning control and self-tuning control based on gain scheduling. The self-tuning approach of quasi-sliding mode control design performs three basic tasks: (i) information gathering of the present process behavior, (ii) control performance criterion optimization, and (iii) adjustment of the controller (Isermann et al., 1992). In this case, the controller parameters are directly identified, because the process model is inserted in the controller design equation.

The realized simulations treat the controller design for analysis of servo behavior. In self-tuning control design, a genetic algorithm optimizes three parameters of controller design. The optimized design parameters are k_1, k_2, and f_0 (see Figure 5.9).

In the self-tuning control based on the gain scheduling approach, a genetic algorithm optimizes three parameters of design for two set point changes. The optimized design parameters are k_1, k_2, and f_0 for samples 1–80, k_1, k_2, and f_0 for samples 81–160, and k_1, k_2, and f_0 for samples 161–240. In this case, the total number of design variables is nine parameters. A same parameter r is utilized for all set points. The relevant parts and implementation of self-tuning quasi-sliding mode control of a canonical genetic algorithm for a case study are described as follows.

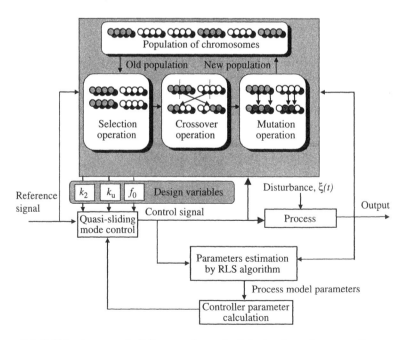

Figure 5.9 Self-tuning quasi-sliding mode control design based on genetic algorithms.

- *Mathematical model*: The discrete mathematical model of a process to be controlled is represented by an ARMA-type model given by $A(q^{-1})y(t) = q^{-d}B(q^{-1})u(t) + \xi(t)$. The changes in the polynomials $A(q^{-1})$ and $B(q^{-1})$ are given by the following conditions:
  - Samples 1 to 80:

$$A(q^{-1}) = 1 - 1.036q^{-1} + 0.2636q^{-2} \quad \text{and} \quad B(q^{-1}) = 0.1387q^{-1} + 0.0889q^{-2}$$

  - Samples 81 to 160:

$$A(q^{-1}) = 1 - 1.976q^{-1} + 0.3477q^{-2} \quad \text{and} \quad B(q^{-1}) = 0.00575q^{-1} + 0.1663q^{-2}$$

  - Samples 161 to 240:

$$A(q^{-1}) = 1 - 0.821q^{-1} - 0.067q^{-2} \quad \text{and} \quad B(q^{-1}) = 0.0275q^{-1} + 0.2663q^{-2}$$

  where the time delay, d, is considered known and equals zero. The noise variance (an uncorrelated random sequence) is set to 0.001 for all simulations.
- *Prediction model*: The prediction model uses $na = nb = 2$. The parameters of polynomials $A(q^{-1})$ and $B(q^{-1})$ are estimated using a recursive least-squares algorithm. An exponential forgetting factor is adopted with values ranging from 0.95 to unity. The estimated parameters are initialized in simulation with $\theta = \{a_1, a_2, b_1, b_2\} = \{0.2, 0.2, 0.2, 0.2\}$. The diagonal of the initial covariance matrix is set to 1000.
- *Reference trajectory*: The desired reference signal is given $y_r(t) = 1$ for samples 1–40, 81–120, and 161–200, and $y_r(t) = 2$ for the samples 41–80, 121–160, and 201–240.
- *Objective function*: The objective function of self-tuning quasi-sliding mode control is given by Equation (4.33), and the control law of Equation (4.34) is obtained by minimization of Equation (4.33).
- *Optimization procedure*: A genetic algorithm is used in the optimization procedure of parameters k_1, k_2, and f_0 of quasi-sliding mode control design. In the sequel, it illustrates the main features of the genetic algorithm employed.

  - *Fitness function*: The fitness function to be maximized is represented by the following equation:

$$fitness = \frac{k_s}{1 + J_{ga}}$$

where the scale coefficient $k_s = 10$, and the performance index is given by $J_{ga} = p[y_r(t) - y(t)]^2 + r[u(t) - u(t-1)]^2$ with $p = 1$.

- *Chromosome structure and coding*: A canonical genetic algorithm is used in this example, and the optimization variables k_1, k_2, and f_0 are coded as real variables. The adopted population size is set to 30 chromosomes, and the search range for each variable is:

$$\begin{cases} 0 < k_1 < 2 \\ 0 < k_2 < 1 \\ 0 < f_0 < 1 \end{cases}$$

- *Selection operator*: A set of individuals from the previous population must be selected for reproduction. The selection operator used in this example is the roulette wheel.
- *Crossover operator*: The crossover operator for this example employs a simple crossover with crossover probability $p_c = 0.80$.
- *Mutation operator*: The mutation operator is a method for introducing new structures into the population. The used mutation probability is $p_m = 0.05$
- *Termination criterion*: The adopted termination criterion is the number of generations equal to 100. In this case, the number of experiments for each design is set to 20.

5.2.3   Simulation results

5.2.3.1   Case study 1: Self-tuning quasi-sliding mode control

This section shows the results obtained with the self-tuning quasi-sliding mode control design. Table 5.2 summarizes the performance and design parameters of a self-tuning controller optimized by genetic algorithms. Figures 5.10 to 5.14 show the performance of best designs for each self-tuning quasi-sliding mode controller.

Table 5.2   Simulation Results for Self-Tuning Quasi-Sliding Mode Control

	Design Parameters				Indices		
r	k_1	k_2	f_0	Fitness	N $\Sigma e(t)^2$ $t = 1$	N $\Sigma \Delta u(t)^2$ $t = 1$	N $\Sigma u(t)^2$ $t = 1$
0.5	0.533	0.008	0.219	0.351	22.478	8.161	387.024
1	0.557	0.1451	0.239	0.325	24.620	5.912	383.718
2	1.349	0.063	0.392	0.276	25.182	5.071	383.043
5	1.998	0.839	0.502	0.212	34.553	2.396	373.394
20	1.968	0.875	1.000	0.125	65.052	0.686	356.765

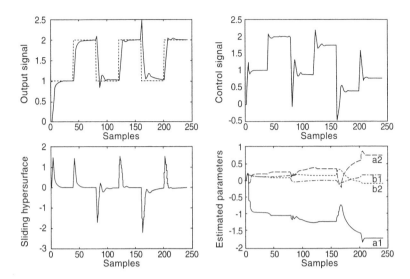

Figure 5.10 Self-tuning quasi-sliding mode control with $r = 0.5$.

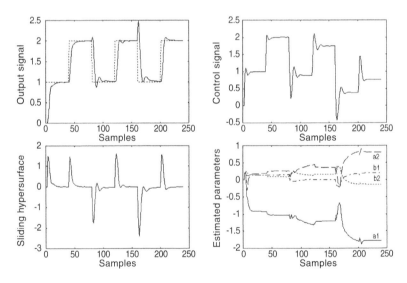

Figure 5.11 Self-tuning quasi-sliding mode control with $r = 1$.

The control designs, presented in Table 5.2, behaved well with a small overshoot without deterioration of the quality of the control signal. However, the performance of controllers was affected by parameter estimation for samples 161–200.

The self-tuning quasi-sliding mode control designs with $r = 0.5$, $r = 1.0$, and $r = 2.0$ presented adequate performance in control of the process. Furthermore, these controller designs obtained fast response, reasonable control activity, and good set point tracking ability.

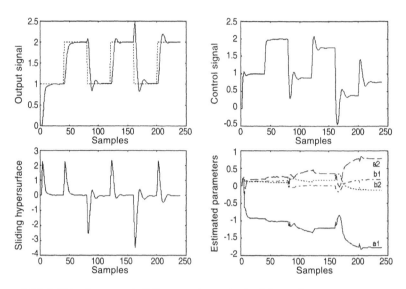

Figure 5.12 Self-tuning quasi-sliding mode control with $r = 2$.

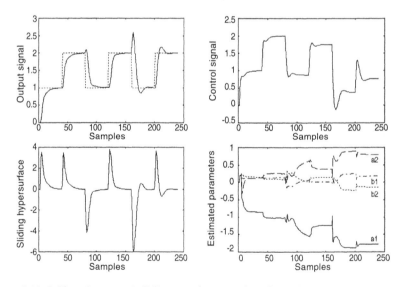

Figure 5.13 Self-tuning quasi-sliding mode control with $r = 5$.

The control designs with $r = 5$ and $r = 20$ presented slow output tracking response. However, these controllers present control signal variance considerably less active than the controller designs with $r = 0.5$, $r = 1.0$, and $r = 2.0$.

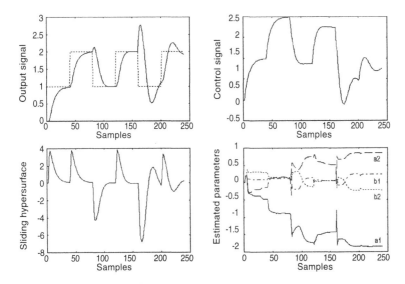

Figure 5.14 Self-tuning quasi-sliding mode control with $r = 20$.

Table 5.3   Simulation Results for Self-Tuning Quasi-Sliding Mode Control with Gain Scheduling

		Design Parameters				Indices		
						N $\Sigma e(t)^2$	N $\Sigma \Delta u(t)^2$	N $\Sigma u(t)^2$
r	Samples	k_1	k_2	f_0	Fitness	$t = 1$	$t = 1$	$t = 1$
0.5	1–80	0.729	0.231	0.247				
	80–160	0.651	0.008	0.231	0.355	26.163	11.693	386.551
	161–240	0.675	0.090	0.216				
1	1–80	0.557	0.110	0.192				
	80–160	0.424	0.020	0.196	0.327	23.890	6.113	381.923
	161–240	0.094	0.043	0.145				
2	1–80	1.184	0.078	0.620				
	80–160	0.871	0.373	0.349	0.272	28.553	4.238	387.652
	161–240	1.051	0.361	0.298				
5	1–80	1.953	0.875	0.780				
	80–160	1.686	0.349	0.576	0.217	35.587	2.235	376.145
	161–240	1.741	0.878	0.447				
20	1–80	0.730	0.067	0.996				
	80–160	1.749	0.749	0.988	0.119	70.383	0.649	358.335
	161–240	1.310	0.988	0.965				

5.2.3.2   Case study 2: Self-tuning quasi-sliding mode control with gain scheduling

This section shows the results obtained with a self-tuning quasi-sliding mode control with gain scheduling. Table 5.3 summarizes the performance and design parameters of self-tuning controller optimized by genetic algorithms.

Figures 5.15 to 5.19 show the performance of each evaluated controller. The fitness evolution versus generations and the convergence of design parameters are presented in Figures 5.20 and 5.21, respectively.

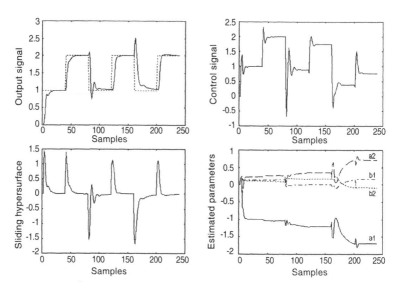

Figure 5.15 Self-tuning quasi-sliding mode control design with gain scheduling ($r = 0.5$)

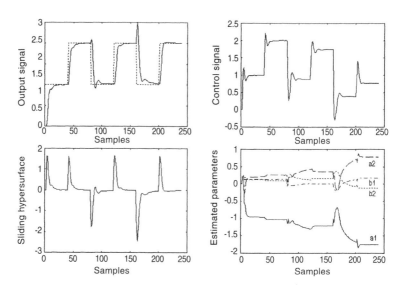

Figure 5.16 Self-tuning quasi-sliding mode control design with gain scheduling ($r = 1.0$).

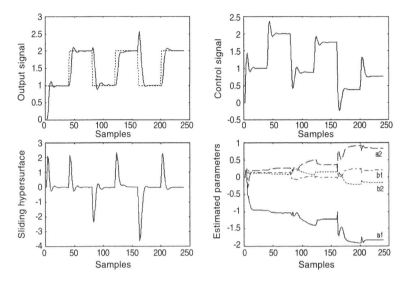

Figure 5.17 Self-tuning quasi-sliding mode control design with gain scheduling ($r = 2.0$).

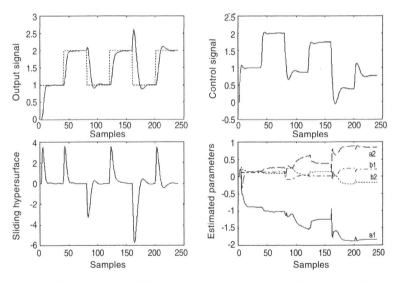

Figure 5.18 Self-tuning quasi-sliding mode control design with gain scheduling ($r = 5.0$).

The best performance of evaluated self-tuning quasi-sliding mode control designs was with $r = 1.0$. The performance of controllers with $r = 5$ and $r = 20$ was significantly inferior to that of the other studied gain scheduling approaches.

Any measures of performance for the simulations were presented in Table 5.3. Based on this table, the best index for sum of square errors was

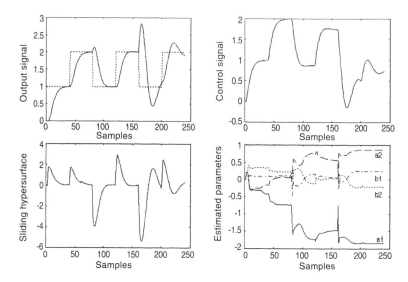

Figure 5.19 Self-tuning quasi-sliding mode control design with gain scheduling ($r = 20.0$).

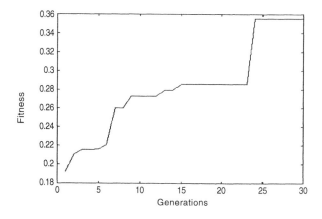

Figure 5.20 Fitness convergence of quasi-sliding mode control GPC design (best chromosome) for $r = 1.0$.

obtained using the gain scheduling design with $r = 1.0$. The design with $r = 20$ presents increasing of sum of square errors and reduced control activity.

The proposed quasi-sliding mode control design with gain scheduling has a large number of tuning parameters in relation to the other analyzed self-tuning controller. This characteristic adds flexibility to tuning automatic procedure based on genetic algorithms, but it can cause problems due to "many degrees of freedom" of the controller. Furthermore, the proper tuning of the controller can be a computationally complex task due to the number of tuning parameters.

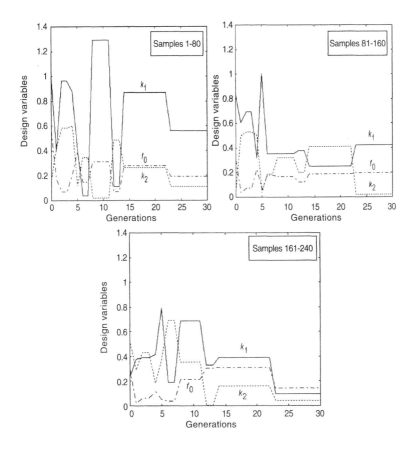

Figure 5.21 Convergence of genetic algorithm for quasi-sliding mode control design with gain scheduling (best chromosome for $r = 1.0$).

5.3   Conclusions

This chapter presented the development of a canonical genetic algorithm to the design of GPC and variable structure systems of quasi-sliding mode type. Furthermore, from the numerical simulation examples, the effectiveness of the proposed control schemes were shown. The utilization of genetic approach avoids the tedious manual trial-and-error procedure, and it presents robustness in the tuning of design parameters.

The proposed two GPC approaches were based on adaptive design without constraints and adaptive design with constraints for the control signal. The optimized design parameters of GPC were the maximum output horizon, N_2, the control horizon, N_u, and the weighted sequence on the control increments, Γ. The obtained simulation results for servo behavior with and without constraints were suitable for GPC design.

However, other conceptions of GPC design can be realized by genetic algorithms, such as: (i) new choices of fitness, (ii) identification of structure and parameters of the mathematical model of process, and (iii) modification of the control law for simulations with nonlinear and constrained processes.

Two design conceptions of quasi-sliding mode control with parameters optimization by genetic algorithms were adopted: (i) self-tuning control and (ii) self-tuning control based on gain scheduling. The optimized design parameters by genetic algorithms are: k_1, k_2, and f_0. The simulation results demonstrated the adequate performance of genetic design of this control type, where the genetic algorithm presents design flexibility for this control design.

Although the basic genetic operators can give satisfactory results in a wide class of complex problems, it is well known that a significant improvement in the algorithm convergence speed can be obtained by defining specialized ad-hoc operators depending on the specific problem. Some improvements can be included in the canonical genetic algorithms used in this chapter for obtaining better results when compared with the standard genetic algorithm. The convergence rates for a local search of genetic algorithms can be improved substantially by utilizing mechanisms of hill-climbing, such as Baldwin effect and Lamarckian learning (Whitley et al., 1994). Other possibilities relevant in GPC design are that the prior knowledge in design be incorporated in the initial population of the genetic algorithm.

The GPC and variable structure systems designs based on genetic optimization can be applied to processes with relatively slow dynamics (chemical and biomedical processes) and long control horizons for off-line (or on-line) design of control laws. The performance of the genetic design is usually superior to that of the manually designed control system in terms of transient response and steady-state error. Therefore, a designer needs to understand the trade-off between control accuracy and computational costs of genetic optimization for a better evaluation of control designs.

The encouraging results presented in this chapter show that further research into different combinations of linear and nonlinear control and optimization strategies may prove beneficial. The aim of future works is to investigate the use of genetic algorithms combined with other computational intelligence methodologies, such as fuzzy systems and neural networks for optimization in multivariable nonlinear model-based predictive control and variable structure control. Furthermore, other relevant studies can be realized, such as: (i) comparative analysis of evolutionary algorithms and sequential quadratic programming, (ii) conceptions of penalty function methods (Gray et al., 1995) for constrained control, (iii) verification of number of floating-point operations of each design, and (iv) automatic adjustment of crossover and mutation probabilities (Srinivas and Patnaik, 1994), because poor choice of the genetic algorithms parameters sometimes resulted in unacceptable behavior of optimization procedures.

References

Gray, G.J. et al., Specification of a control system fitness function using constraints for genetic algorithm based design methods, *Proceedings of the 1st IEE/IEEE GALESIA*, Sheffield, United Kingdom, pp. 530–535, 1995.

Isermann, R., Lachmann, K.-H., and Matko, D., *Adaptive control systems*, Prentice Hall, Englewood Cliffs, New Jersey, 1992.

Srinivas, M. and Patnaik, L.M., Adaptive probabilities of crossover and mutation in genetic algorithms, *IEEE Transactions on Systems and Cybernetics*, 24, 4, 656–667, 1994.

Whitley, D., Gordon, V.S., and Mathias, K., Lamarck evolution, the Baldwin effect and function optimization. Parallel problem solving from nature, *Lecture Notes in Computer Science*, v. 866, Springer-Verlag, Berlin, pp. 6–15, 1994.

chapter six

Tuning fuzzy logic controllers for robust control system design

6.1   Introduction

The limitations of conventional controllers for application to complicated, dynamical systems have motivated research into "intelligent" control systems. A popular technique is fuzzy control, in which expert knowledge can be incorporated into the design. An introduction to fuzzy logic and fuzzy control is provided in Appendix A.

Fuzzy logic initially proved attractive because of its capability to mimic human intelligence, and thus, the concept of fuzzy systems was soon associated with many practical applications. The real boom of fuzzy technology started in the mid 1980s to early 1990s in Japan, where a whole range of industrial products based on approximate reasoning was developed. Well-known examples are the transport system control for the Sendai Subway, the control of robots at Fuji Electric, and a TV set based on fuzzy technology built by Sony. Although Japan can be credited with the birth of the fuzzy logic industry, elsewhere, the new technology gradually gained acceptance.

Alongside this historical perspective of fuzzy systems is the role played by this type of expert system in science and technology. Fuzzy systems were not intended to replace conventional methods that had well-established theories and many successful applications, but rather to complement them. The ultimate purpose of fuzzy systems was to tackle problems that did not have feasible or straightforward solutions using traditional methods and to make use of human experience. In industrial examples such as the one studied here, an appeal of fuzzy control is its ability to act as a sophisticated nonlinear controller founded on some simple rules, readily understood and proposed by practicing industrial engineers. This chapter describes how genetic algorithms can assist in the design of fuzzy controllers and demonstrates this through a case study example.

In fuzzy control, a genetic algorithm can be used, off-line, for the following:

- Tuning of the membership functions
- Elicitation of the rule base in addition to tuning

Method (1) is demonstrated in this chapter. Practitioners of this approach tend to argue that the form of the rules is likely to be known *a priori*, and that most uncertainty lies in the development of the associated membership functions. Use of a static rule base also reduces the necessary level of computational complexity, which may be another reason for the popularity of this approach.

Section 6.2 introduces the subject of fuzzy control and provides useful background literature references. Aspects of genetic tuning are introduced and discussed in Section 6.3. The control of gas turbine engines (GTEs) forms the basis of the example study used to demonstrate the application of genetic tuning. In Section 6.4, following an overview of GTEs and an introduction to their different types, the nature of the control problem is explained. In Section 6.5, fuzzy controllers are first derived using heuristic methods and then refined with the aid of genetic tuning. Section 6.6 concludes by explaining how the versatility of GAs enables more complex problems to be addressed, which build upon the simple example described in Section 6.5. Other applications of GAs in fuzzy control are also described.

6.2   Fuzzy control

When dealing with multivariable, nonlinear systems, or when the information about the process is scarce or unavailable, traditional systems theory may not be able to provide efficient control. Through local linearization, the true character of the system may be misrepresented, and the control method may be inadequate. If the system can be approximated with local linear functions, the theory behind the control system is valid over a limited scope, and the system is subject to restrictive assumptions. Moreover, when the analysis and the design of a system are performed using linear control theory, an evaluation of the resulting controller on the original nonlinear system is normally required, and a redesign process may be necessary (Kuo, 1995). To overcome these deficiencies of classical control and also to incorporate human skills into the control strategy, fuzzy systems may be employed as "expert" linguistic controllers. Driankov et al. (1993) and Passino and Yurkovich (1998) provide introductions to the subject of fuzzy control, which cover salient aspects of this discipline.

The competence of fuzzy systems in control is demonstrated by the vast number of successful applications in diverse topics in control. In Kosaki et al. (1997), for example, fuzzy control is applied to a magnetic bearing system. TSK-TSK and Mamdani systems are discussed in Section 6.3 — a Takgi-Sugeno-Kang (TSK) fuzzy controller — is designed by means of Lyapunov stability condition, and the performance of the resulting system is assessed through computer simulation. A similar approach is used in Leung et al. (1998). Here, Lyapunov theory is used for the design of a fuzzy controller capable of stabilizing a nonlinear mass–spring–damper system with unknown parameters. In fact, the issue of designing stable fuzzy controllers using the Lyapunov method is widely studied (see Palm et al., 1997).

Mamdani systems are equally applicable and used in control problems. For instance, Oriolo et al. (1999) use Mamdani fuzzy mappings for the real-time control of mobile robots. In this case, simple linguistic rules dictate the robot motion in a certain environment so that collisions with obstacles implanted in the robot neighborhood are avoided.

These examples are but a few in a large and growing area of research. The first major survey of fuzzy logic control and its applications was undertaken by Lee (1990). More recently, Isermann (1998) summarized and published some of the latest achievements in this area.

Before concluding this section, it is also worth mentioning that fuzzy logic has proved very useful for modeling systems that are subject to uncertainty, are ill-defined, or possess unmodeled dynamics. Hellendoorn and Driankov (1997) have collected and published some of the most important papers in this area and also present an informative overview of fuzzy model identification techniques. While our example study in this chapter focuses on fuzzy control, GAs have also proved effective in tuning fuzzy models.

6.3   Genetic tuning of fuzzy control systems

A fuzzy system is comprised of three major components:

1. *Fuzzification*: An input interface, which realizes the conversion of crisp inputs into fuzzy inputs by means of fuzzy sets.
2. *Inference engine*: A collection of fuzzy rules in the form "if <antecedent> then <conclusion>," where knowledge about the problem is acquired. A mechanism of inference then deals with the fuzzy rules in order to generate fuzzy conclusions (consequents) from the fuzzy premises (antecedents).
3. *Defuzzification*: An output interface, which transforms fuzzy outputs into crisp outputs.

The general structure of a linguistic fuzzy system is shown in Figure A.15.

Fuzzy systems in which both the antecedent and the consequent rule part are expressed through fuzzy predicates are known as Mamdani or linguistic fuzzy systems. When fuzzy systems embody fuzzy and nonfuzzy descriptions of the variables of interest, the systems are referred to as Takagi–Sugeno–Kang (TSK) fuzzy systems. The latter system emerged as an alternative to Mamdani's linguistic formulation, and the idea was to alter the rule structure so that qualitative and quantitative knowledge can be equally incorporated into the knowledge base. This dual representation is realized by expressing the consequence of the rule in an analytical form, while the fuzzy antecedent part of the rule retains the linguistic formulation from the Mamdani case.

The defuzzification process is not required for TSK systems, as the outputs are represented in the rule base as mathematical expressions (usually regressions), which produce directly crisp data. The overall output of the TSK model is calculated as a weighted sum of the outputs resulting from all the contributor rules.

The Mamdani formulation is adopted in the example study in this chapter.

A contributing factor to the success of fuzzy technology is the transparency exhibited by these systems. The human operator can understand and interpret the information contained in the rule base and can relate this information back to the physical proprieties of the process. All of these characteristics are the result of an accumulation of factors, which include:

- The linguistic character of the system and the generalization of information by means of fuzzy sets
- The use of heuristic rules to affect the system behavior
- The interpretation given to the rules contained in the knowledge base through the following:
  - The size and number of partitions of the universe of discourse
  - The shape of individual membership functions
  - The type of reasoning mechanism applied to the rule base (the inference engine)

Fuzzy systems are rapidly embraced by many industrial environments, as they are not only flexible and structurally simple, but also easy to understand and cheap to implement.

This transparency does not extend, however, to parameters defining the precise details of the membership functions, and selection of these is a prime candidate for the use of genetic search (*genetic tuning*).

In a fuzzy system, the main tuning variables are as follows:

- *The scaling factors* (if used) — similar to gain tuning in PID controllers
- *The position of the membership functions* — a few options are available here, one or more parameters can be tuned for each function. For all types of membership function, the only tuning parameter could be the crossover point (the intersection points between adjacent membership functions). For triangles or trapeziums, for instance, only the center of gravity can be encoded and used in the optimization process. Conversely, for these cases, the three (four) vertices can be also encoded, if applicable.
- *The rule base* — the entire rule structure, or only the consequence, typically through numerical identifiers

The number of variables to be tuned is determined by the designer, according to the knowledge of the problem and the rigor with which the search should be conducted (i.e., coarse or fine tuning).

6.4 Gas turbine engine control

The gas turbine engine will be the vehicle for our example study. It is a good example of an industrial nonlinear multivariable control problem and has been much studied in the literature. This section describes some of the features of gas turbine engines and the nature of the control problem to be studied.

6.4.1 Gas turbine engines — an overview

Gas turbine engines (GTEs) are internal combustion heat engines (machines that convert heat energy into mechanical energy), which use air as working fluid in order to produce a propulsive jet. A GTE is made up of three main components: a compressor and a turbine placed on a common shaft, and a combustion chamber (Figure 6.1).

The role of the GTE in the aircraft is to generate thrust, by imparting momentum to the fluid passing through it (Hill and Peterson, 1992). This means that the working fluid should be expanded in order to produce a propulsive jet. Hence, the air entering the engine body is compressed by the compressor unit and, consequently, delivered to the turbines, at higher pressure. In order to increase the energy of the working fluid prior to expansion through the turbine, fuel is mixed with the compressed air and ignited at virtually constant pressure, in a combustion chamber. The burning of the air–fuel mixture results in a rise in temperature and, thus, an expansion of the gases. The power developed in this manner is sufficient to actuate the turbine and also to create a propulsive jet, or thrust.

GTE compressors may experience *surge*, a destructive phenomenon that can cause excessive aerodynamic pulsations, which are transmitted through

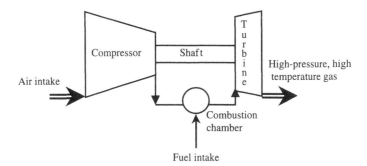

Figure 6.1 A schematic gas turbine engine. (From Bica, B., Ph.D. thesis, University of Sheffield, U.K., 2000. With permission.)

the whole machine and must be avoided at all costs. Over the entire operational range of the engine, a *surge line* is normally defined and used as a measure of aerodynamic stability. The surge line is made up of a number of individual surge points, corresponding to different engine speeds, which span the entire working space.

6.4.2   GTE types

Engine types can be classified according to the number of compressor–turbine pairs (or *spools*) they employ. The three main groups are *single-spool*, *twin-spool*, or *three-spool* engines. The twin-spool structure is the most commonly used. In these engines, the compressor is divided in two separate units (Figure 6.2), each of these achieving a different level of air compression. The first compressor absorbs the atmospheric air and raises its pressure by a certain extent — this is the low-pressure (LP) compressor or *fan*. The second compressor, which is termed the high-pressure (HP) compressor, increases the air pressure still further prior to reaching the combustion chamber. A variable inlet guide vane (IGV) is used to match the air from the fan to the HP compressor characteristics. Each compressor is driven by a turbine mounted on the same shaft.

The engine type employed in this case study is an example of a *twin-spool turbofan* engine, which is one of the most common types of GTE used for aircraft propulsion. In a turbofan engine, a portion of the airflow bypasses the engine core and is then mixed with the combusted gas from the turbine exit before being ejected through the jet pipe and nozzle area (NOZZ) to produce thrust. The thrust generated in this manner will thus have two components: a main hot thrust component coming from the engine core, and a cold stream (or fan) thrust component, resulting from the bypass flow. The bypass duct was designed with the aim of reducing the overall jet velocity

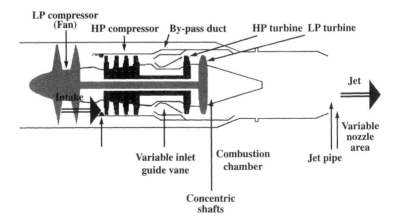

Figure 6.2 The mechanical layout of a twin-spool engine. (From Bica, B., Ph.D. thesis, University of Sheffield, U.K., 2000. With permission.)

by allowing cold air to be added to the main hot jet, reducing its temperature. (The decrease in the propulsive speed due to bypass air leads to better propulsive efficiency, lower noise levels, and improved specific fuel consumption.) Figure 6.2 shows the mechanical layout of a typical twin-spool gas turbine engine.

6.4.3 *The GTE control problem*

From a control perspective, a GTE is a complex plant, subject to stringent constraints and strict performance requirements. Over its entire flight envelope, the engine is required to meet a number of performance specifications, while maintaining stability and safe operation with minimum overall cost. The control problem is further complicated by cross-coupling between different engine parameters, where a variation in one will disturb other control variables. The inherent nonlinear dynamics, added to the multivariable nature of the engine, augment the difficulty of the control problem.

The characteristics of operation of a fixed-cycle gas turbine engine, such as specific thrust and specific fuel consumption, are fundamental to engine design. The design, thus, becomes a compromise between meeting the conflicting requirements for performance at different points in the flight envelope and achieving low life-cycle costs, while maintaining structural integrity. However, variable geometry components, such as the inlet guide vanes and nozzle area, may be used to optimize the engine cycle over a range of flight conditions with regard to thrust, specific fuel consumption, and engine life, assisting in the reduction of life-cycle costs.

Dry-engine control of a conventional engine is normally based on a single closed-loop of fuel flow for thrust rating, engine idle and maximum limiting, and acceleration control. The closed-loop control concept provides accuracy and repeatability of defined engine parameters under all operating conditions and automatically compensates for the effects of engine and fuel system degradation. Figure 6.3 shows the baseline configuration for a typical engine controller block. A nonlinear thermodynamic model of the engine (realized in SIMULINK), with inputs for fuel flow (WFE), HP inlet guide vane angle (IGV), and exhaust nozzle area (NOZZ), is used to simulate dynamic behavior. Further inputs for flight conditions (altitude, Mach number, and temperature) allow the engine operation to be simulated over the full flight envelope. Sensors provided from the engine outputs are high- and low-pressure spool speeds (NH and NL), bypass duct Mach number (DPUP), and turbine and jet pipe temperatures (TBT and JPT). Other outputs, such as the (fan) low-pressure surge margin (LPSM) and gross thrust (XGN), are calculated directly from internal engine parameters. For the purposes of this study, we have two controllers, fuel flow (WFE) and nozzle area (NOZZ), based around a conventional proportional plus integral (PI) controller structure (see Figure 6.3).

A single input, NHDem, derived from the pilot's lever angle, is used to determine the thrust setting. The WFE controller uses this and the measured

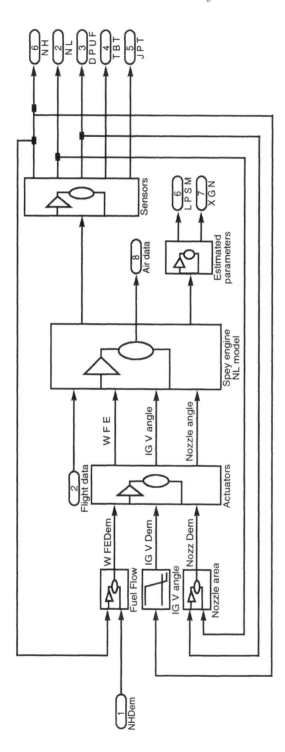

Figure 6.3   Conventional GTE control sceme. (From Bica, B., Ph.D. thesis, University of Sheffield, U.K., 2000. With permission.)

HP compressor speed, NH, to determine the required fuel flow demand. To protect the engine from overacceleration, NHDem is rate limited. A third input, air data, is required to correct the NH value for changes in flight conditions, i.e., temperature and pressure, so that the controller can operate over the full-flight envelope. As the controller is also required to operate over a range of engine conditions, such as idle, cruise, and full power, the fuel flow controller uses gain schedules to accommodate these nonlinearities in the system dynamics.

For the NOZZ controller, the demand input is NL, the fan speed, which is mapped to a desired value of DPUP. A measurement of DPUP is the second input, and a comparison of the two is used to determine the position of the variable geometry components in the jet pipe and, hence, the nozzle area. Adjustment of NOZZ may be used to alter the pressure distributions in the engine and thus to trim the LSPM or thrust level. As for the PI fuel flow controller, gain schedules are used to cover the range of operating conditions. (However, we are only considering a single operating point in this example.)

In recent years, a considerable amount of research has been directed toward the design of controllers for GTEs in an attempt to improve performance and, at the same time, allow reduced production costs. In the next section, an example investigates the use of fuzzy systems for the gas turbine engine control and looks at the capability of the proposed control configuration to improve performance. Genetic algorithms, as tools for search and optimization, assist the design process.

6.5 Fuzzy control system design — example study

6.5.1 Problem formulation

Here, we demonstrate, through a simple example, the feasibility of fuzzy controllers for GTEs, using a genetic algorithm to tune the desired fuzzy control law. For simplicity, a single 85% high-pressure spool speed (NH) operating point is considered at sea level static conditions. The control options considered are the replacement of the PI-based control loops for the WFE and NOZZ loops with Mamdani fuzzy equivalents. As we have seen, the control of the fuel loop is based on measurement of NH, while the nozzle area loop uses measurements of the bypass duct Mach number (DPUP) and fan speed (NL). For this particular operation, the system is required to meet the following design objectives:

1. 70% NH rise time ≤1.4 sec
2. 10% NH settling time ≤1.8 sec
3. XGN ≥34.12KN
4. JPT 680 K
5. LPSM ≥6.6%

where objectives (1) and (2) are in response to a change in high-pressure spool speed demand of 85 to 90% and represent typical dynamic performance requirements for a military engine. Engine thrust, XGN, is the main control parameter to be optimized; recall that it is not measured, rather it is derived from internal engine parameters. JPT is the maximum temperature of the jet pipe and is employed as a measure of thermodynamic stress. A lower value of JPT indicates less stress and, therefore, a longer engine life. Finally, LPSM is the fan surge margin and is employed as a measure of aerodynamic stability and, hence, safety; again it is not measured, instead it is also derived from internal engine parameters.

The design process considers the replacement of the PI WFE and NOZZ controllers simultaneously. Suitable fuzzy structures are found by analyzing the shapes and the domains of the controllers' inputs and outputs, thereby formulating linguistic rules that describe their behavior. The resulting hand-tuned fuzzy controllers should be able to replicate accurately the input-output behavior of the engine's PI controllers. In this study, then, first fuzzy controllers are obtained via a heuristic approach (Section 6.5.2), and next the controllers are tuned using a GA search technique (Section 6.5.3).

6.5.2   Heuristic design of the fuzzy controllers

The design objective is the construction of fuzzy logic mappings that emulate the input-output behavior of the baseline PI controllers. Hence, the fuzzy partitions and rule-base are derived by observing the initial relationships between the inputs and outputs of each engine controller. The first design step is to set the limits of the universe of discourse for each fuzzy parameter. In the original model, for example, NHDem varies within the interval [85, 92] (Figure 6.4). The fuzzy universe of discourse is defined so that it includes this range with a certain tolerance in either side of the real axis: [80, 95] (Figure 6.5). In general, if $\{a, b\}$ are the limits of the original variable, then the fuzzy universe of discourse for the corresponding fuzzy variable has the form: $\{a - tol_a.a, b + tol_b.b\}$. Here tol_a and tol_b represent the lower/upper interval tolerance, expressed as a percentage.

The shapes of the WFE input-output variables [Figure 6.4(a)] are relatively simple and follow a similar pattern. These observations suggest that three membership functions for each variable should be sufficient to build the fuzzy system. The membership functions are positioned intuitively at this stage of the design process (Figure 6.5).

In the WFE controller, the NH demand input is rate limited to provide protection from overacceleration and, thus, has a slower dynamic response than the measurement of NH. This remark, added to observations of the signals characteristic, is used to formulate the membership functions and the rule-base (Table 6.1). For example, in Table 6.1 and Figure 6.5, at the extremes of the maneuver, when both inputs are either low or high, the fuel flow should be low or high, accordingly.

Likewise, the fuzzy nozzle area controller is constructed from observation of the input-output relationship of the PI controller (Figure 6.6 and Table 6.2). Additionally, it may be noted that the number of rules required for

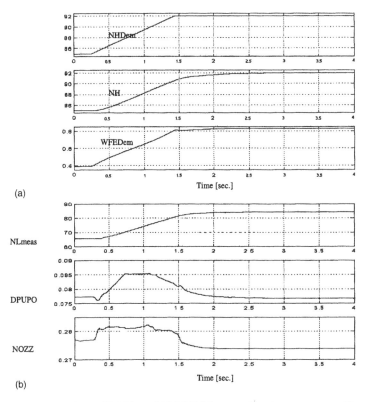

Figure 6.4 The digital PI WFE (a) and NOZZ (b) controllers' parameters. (From Bica, B., Ph.D. thesis, University of Sheffield, U.K., 2000. With permission.)

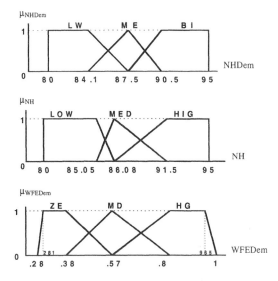

Figure 6.5 The membership functions for the fuzzy WFE controller parameters. (From Bica, B., Ph.D. thesis, University of Sheffield, U.K., 2000. With permission.)

Table 6.1   Rule base for the WFE controller

NH/NHDem	LW	ME	BI
Low	ZE	ZE	MD
Medium	MD	MD	MD
High	MD	HG	HG

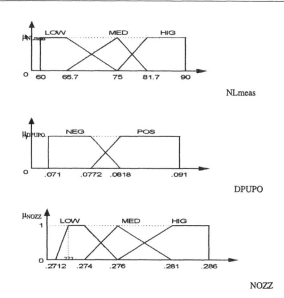

Figure 6.6 The membership functions for the fuzzy NOZZ controller parameters. (From Bica, B., Ph.D. thesis, University of Sheffield, U.K., 2000. With permission.)

Table 6.2   Rule base for the NOZZ controller

DPUPO/NLmeas	LOW	MED	HIG
NEG	LOW	LOW	HIG
POS	MED	MED	LOW

adequate control is small, as the maneuver is only around a small area of the full flight envelope. The fuzzy controllers are not, therefore, necessarily designed to implement a PI-like controller. These observations apply to both fuzzy controllers.

Figure 6.7(a) and (b) depict the outputs of the fuzzy WFE and NOZZ controllers, which are indicated with a dashed line, plotted against the original PI controllers outputs, represented by a continuous line. The results presented in Figure 6.7 indicate that the empirically derived fuzzy controllers are capable of mimicking the original controllers' outputs. Furthermore, the responses of the fuzzy systems are smoother in comparison with the PI controller's outputs. The steady-state error of each output is zero, indicating a good system response.

After adequate fuzzy structures are obtained, their ability to control the plant is verified on the SIMULINK nonlinear thermodynamic engine model. The performance of the PI and fuzzy controllers is compared in Figure 6.8

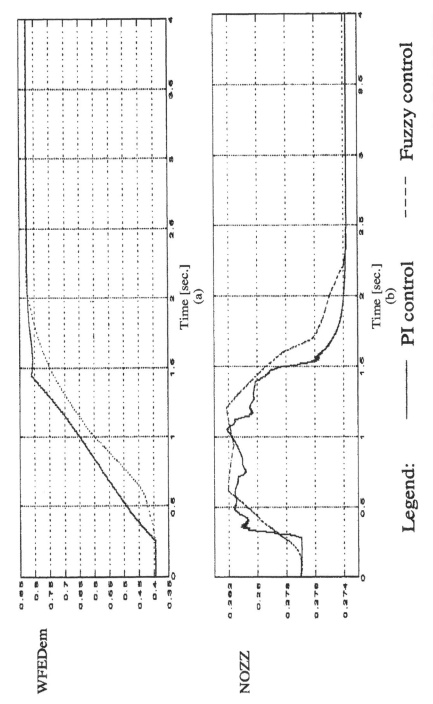

Figure 6.7 The PI and fuzzy WFE (a) and NOZZ (b) controllers' outputs. (From Bica, B., Ph.D. thesis, University of Sheffield, U.K., 2000. With permission.)

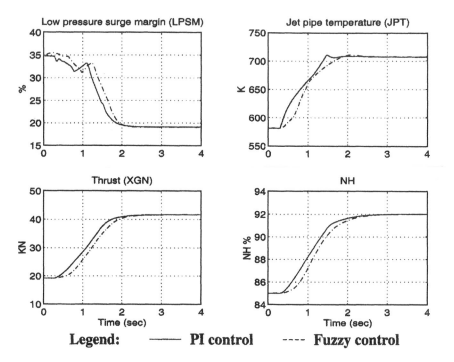

Legend: ——— **PI control**   - - - - **Fuzzy control**

Figure 6.8 Engine responses with PI and fuzzy controllers. (From Bica, B., Ph.D. thesis, University of Sheffield, U.K., 2000. With permission.)

for the NH step response. It can be observed that the fuzzy controller, denoted by a dashed line, offers a satisfactory level of control for this maneuver, meeting the specified design requirements, while offering marginally improved LPSM and JPT characteristics. This means that in the fuzzy controller case, the LPSM is slightly increased, while the JPT has a lower average value than in the PI controller case.

6.5.3   GA tuning of the fuzzy controllers

It is to be expected that, following such a heuristic approach to fuzzy control design, it will be commonplace to want to apply some tuning to the selected parameters. Genetic algorithms are natural search engines to select for this task. Thus, having designed fuzzy controllers for the operating point of interest that satisfactorily approximate the original PI controllers' responses, the rule base and membership functions are now tuned.

The encoding of each fuzzy rule for use in the GA is made up of a coding of the linguistic consequent part of the rule and the position of the membership functions. An individual encodes the parameters of both controllers, and they are, therefore, composed of a set of 15 rules (see Tables 6.1 and 6.2) represented by integer identifiers and six output membership functions represented with real values for their coordinates in the parameter space(Figure 6.9).

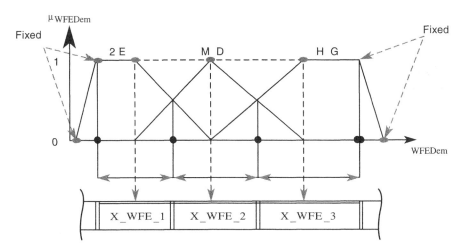

Figure 6.9 Chromosome for the membership function representation. (From Bica, B., Ph.D. thesis, University of Sheffield, U.K., 2000. With permission.)

X_WFE_i denotes the encoded coordinates in Figure 6.9. The limits within which each coordinate is allowed to vary are defined by the crossover points of the adjacent membership function and are also indicated in the figure. The universe of discourse of each parameter is lower and upper bounded in order to limit the size of the search space and to ensure that the membership functions cover all possible values of the control parameters. As a result, the lowest and the highest values of the coordinates do not change during tuning, as indicated.

Simple GAs with populations of 40 real-valued individuals were employed. The individuals were encoded as described above with the addition of constraints on the relative positions of the membership functions being included directly in the representation. The recombination operator was applied with a probability of 0.9 during recombination, and mutation was then employed with a probability of 0.1. The single objective to be maximized was the system thrust resulting from the engine simulation. Thus, this is a reasonably compute-intensive approach, requiring one simulation (using SIMULINK) to evaluate the objective function, *thrust*, for each individual used in the search. Because the GA is a stochastic search method, a series of seven runs were performed for result analysis purposes.

Figure 6.10 illustrates the behavior of the gas turbine engine with GA-tuned fuzzy controllers. The thick line indicates the system performance with the original PI controllers and the alternative GA-tuned fuzzy controllers with fine lines. Each fine line curve indicates a solution resulting from one GA run.

Comparing the results of Figure 6.8 (heuristic design) with those of Figure 6.10 (GA-tuned design), it can be seen that the target objective, *thrust*, is greatly improved in the GA-tuned fuzzy control approach. From the optimization point of view, the goal of maximizing engine thrust is therefore

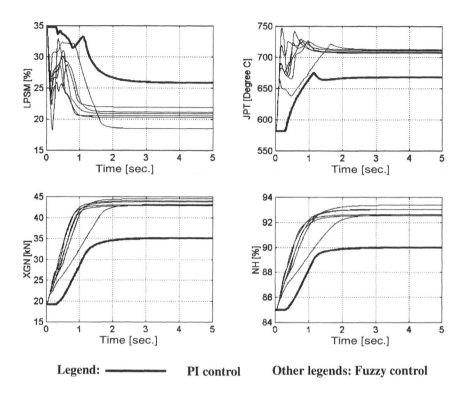

Legend: ━━━━━━━━ **PI control**   **Other legends: Fuzzy control**

Figure 6.10 Performance of the engine with the GA-tuned fuzzy controllers. (From Bica, B., Ph.D. thesis, University of Sheffield, U.K., 2000. With permission.)

achieved. The considerable difference in engine thrust in the hand-tuned (heuristic) and GA-tuned approaches, as seen in Figure 6.10, is attributed to different settings of nozzle area. A reduction in the nozzle area improves thrust but has the shortcoming of decreasing the surge margin and increasing the jet pipe temperature. Therefore, lower surge margins and higher jet pipe temperatures are the price for improving the engine thrust. This observation also points to the fact that conflicts exist between different control objectives; an improvement in one objective leads to degradation in other objectives. This finding suggests that the inherent trade-offs in the GTE should be taken into account in the design and optimization processes. Moreover, a simultaneous consideration of various conflicting objectives could offer the engineer better insight into the nature of these trade-offs and the choices between optimal design solutions. In Chapter 9, we introduce the multiobjective genetic algorithm, which can be used to address such multiple objectives.

6.6   Applications of GAs for fuzzy control

The example used in this chapter demonstrated an approach to the design of fuzzy controllers for a gas turbine engine based on plant knowledge and observations of the plant's behavior for a single-operating point scenario.

Using pure heuristics as guidelines, the original PI controllers have been successfully replaced by their fuzzy equivalents. However, control and modeling approaches often require optimization in order to achieve desired levels of performance. Among many global optimization methods, GAs stand out as powerful, flexible techniques, applicable to a wide range of engineering problems. For this reason, a genetic algorithm optimization has been used to fine-tune the parameters of a known (hand-tuned) fuzzy equivalent of the engine PI controllers. Using the GA approach, thrust maximization (the primary design and optimization aim) has been achieved. However, a number of other critical parameters, such as surge margin or jet pipe temperature, have not been taken into account in the optimization process. Bica et al. (1998) have demonstrated the benefits of GA-based *multiobjective* tuning of fuzzy controllers for a GTE.

Fuzzy systems can generate smooth multidimensional input-output mappings and thus be very good interpolators, a fact that explains their frequent use as gain schedulers. Fuzzy gain scheduling should, therefore, be seen as a major part of fuzzy control and one of the main contributors to the success of this discipline. For example, Bica et al. (2000), have extended the single-operating point approach of Section 6.5 to produce fuzzy scheduled controllers that operate over a wide flight envelope.

Now, to conclude, we turn to some non-GTE applications of fuzzy control. Karr (1991) was among the first to use GAs for tuning fuzzy systems. He proposed an approach for tuning off-line the system's membership functions, arguing that there was no need to include the rules in the optimization process. In his view, the rules derived empirically by the human expert can be valid over a wide range of operating conditions.

Ke et al. (1998) applied a "hierarchical" GA to tune a fuzzy logic controller for a solar power plant. Essentially, this algorithm is a variant of the variable-length GA. This approach allows the selection of an optimal number of membership functions by omitting those fuzzy partitions that are found to be unsatisfactory in the evaluation process. More precisely, the GA encodes the positions of the membership functions in a real-valued chromosome, whereas binary encoding is used to symbolize the membership function identifiers. A binary value of 1 of the identifier indicates the presence of the corresponding function from the fuzzy universe of discourse, and a binary value of 0 indicates its absence. The binary identifiers and the real value chromosomes are concatenated to create an individual in the GA, which is then subject to genetic transformations. After the exclusion of redundant membership functions, a "recovery" procedure is performed in order to fill the gaps created in the universes of discourse. To achieve this, the valid functions are rescaled to cover all the undefined regions.

Another type of variable-length GA was proposed by Leitch and Probert (1998) for the derivation of the fuzzy partitions and control rules to operate a cart-pole system. The proposed GA was based on a context-dependent coding (CDC), which involved the utilization of identifiers (context switches) to break binary chromosomes into different sized sections, each representing a fuzzy variable. The authors reportedly obtained better results compared with other GA encoding schemes (fixed-length, real-valued).

Last, in a more exotic application (Nawa et al., 1997), a pseudo-bacterial GA was used to build a fuzzy logic controller for a semiactive suspension system. The algorithm attempts to mimic the process of bacterial recombination, where the "best" characteristics of one bacterium can be spread among the entire bacteria population before being subject to crossover. The approach was allegedly successful in the acquisition of fuzzy rules for the chosen application.

References

Bica, B., Ph.D. thesis, University of Sheffield, U.K., 2000.

Bica, B. et al., Multiobjective design of a fuzzy controller for a gas turbine aero-engine, *Proceedings of UKACC International Conference on CONTROL '98*, Swansea, United Kingdom, pp. 901–906, 1998.

Bica, B., Chipperfield, A.J., and Fleming, P.J., Towards fuzzy gain scheduling for gas turbine aero-engine systems: A multiobjective approach, *Proceedings of IEEE International Conference on Industrial Technology 2000 (ICIT 2000)*, Goa, India, pp. 81–86, 2000.

Driankov, D., Hellendoorn, M., and Reinfrank, M., *An Introduction to Fuzzy Control*, Springer-Verlag, Heidelberg, 1993.

Hellendoorn, M. and Driankov, D., Eds., *Fuzzy Model Identification*, Springer-Verlag, Heidelberg, 1997.

Hill, P. and Peterson, C., *Mechanics and Thermodynamics of Propulsion*, Addison-Wesley, Reading, Massachusetts, 1992.

Isermann, R., On fuzzy logic applications for automatic control, supervision, and fault detection, *IEEE Transactions on Systems Man and Cybernetics*, 28, 2, 221–235, 1998.

Karr, C.L., Design of an adaptive fuzzy logic controller using a genetic algorithm, in *Genetic Algorithms: Proceedings of the Fourth International Conference*, Belew, R. and Brooker, L., Eds., Morgan Kaufmann, San Mateo, California, 1991.

Ke, J.Y. et al., Hierarchical genetic fuzzy controller for a solar power plant, *Proceedings of the IEEE International Symposium on Industrial Electronics*, Vol. 2, pp. 584–588, 1998.

Kosaki, T., Sano, M., and Tanaka K., Model-based fuzzy control system design for magnetic bearings, *Proceedings of the Sixth IEEE International Conference on Fuzzy Systems*, New York, Vol. 2, pp. 895–899, 1997.

Kuo, B.C., *Automatic Control Systems*, Prentice Hall, Englewood Cliffs, New Jersey, 1995.

Lee, C.C., Fuzzy logic in control systems: Fuzzy logic controller — Parts I & II, *IEEE Transactions on Systems, Man and Cybernetics*, 20, 2, 404–435, 1990.

Leitch, D. and Probert, P.J., New techniques for genetic development of a class of fuzzy controllers", *IEEE Transactions on Systems, Man, and Cybernetics*, 28, 1, 112–122, 1998.

Leung, F.H.F., Lam, H.K., and Tam, P.K.S., Adaptive control of multivariable fuzzy systems with unknown parameters, *Proceedings of the 24th Annual Conference of the IEEE Industrial Electronics Society, IECON'98*, New York, Vol. 3, pp. 1758–1761, 1998.

Nawa, N.E. et al., Fuzzy logic controllers generated by pseudo-bacterial genetic algorithm with adaptive operator, *IEEE International Conference of Neural Networks*, Vol. 4, pp. 2408–2413, 1997.

Oriolo, G., Ulivi, G., and Vendittelli, M., Real-time map building and navigation for autonomous robots in unknown environments, *IEEE Transactions on Systems, Man and Cybernetics*, 28, 3, 316–333, 1999.

Palm, R., Driankov, D., and Hellendoorn, H., *Model Based Fuzzy Control*, Springer-Verlag, Heidelberg, 1997.

Passino, K.M. and Yurkovich, S.S., *Fuzzy Control*, Addison-Wesley, Reading, Massachusetts, 1998.

chapter seven

GA-fuzzy hierarchical control design approach*

7.1 Introduction

The real world environment includes parameters, which are often difficult or impossible to accurately represent mathematically. Robotic agents and systems, which operate in the real world, are often corrupted by unstructured, noisy, changing, and known or unknown environmental parameters. A vivid example of this scenario is the safe wandering of the pathfinder on the surface of Mars in July 1997. Consequently, a successful design for a controller should demonstrate the robustness and the adaptation capability to properly compensate for these changing environmental conditions. This is especially prudent in space applications, where the robot must autonomously, and with little or no communication with earth, face the environment and perform complicated multifaceted tasks. This is also a significant factor in robotic waste handling, where the robot operates in unstructured and unknown environments of a waste tank or a waste disposal site.

Control systems based on fuzzy logic have exhibited robustness to unstructured and noisy environments. Fuzzy logic allows for mathematical expression of human knowledge, thereby allowing to embed human knowledge and intuition into an autonomous control system and bypass the complexity involved in some nonfuzzy approaches. However, there are two problems in conventional fuzzy reasoning. One is the lack of a definitive method to determine the membership functions and rules, and the second is the lack of a learning ability on the part of the fuzzy controller.

In order to incorporate a learning ability into fuzzy systems, various optimization (or learning) paradigms such as neural networks, genetic programming, and genetic algorithms may be used (Akbarzadeh and Jamshidi, 1997). Using these optimization techniques, the robotic agent can

* This work is based on Akbarzadeh, M.-R. and Jamshidi, M., *International Journal on Intelligent Automation and Soft Computing*, 3, 1, 77-78, 1997. This work was supported, in part, by NASA-Ames Grants numbers NAG 2-1196 and NAG 2-1480 and by Waste Education and Research Consortium (DOE) under award # WERC/NMSU/DOE Amd 35.

autonomously determine parameters of membership functions and rules. Genetic algorithms are optimization search routines modeled after nature's evolutionary process. Genetic algorithms have demonstrated the coding capability to represent parameters of fuzzy knowledge domains such as fuzzy rule sets and membership functions. In fact, the transformation (the interpretation function) between the fuzzy knowledge domain (phenotype) and the GA coded domain (genotype) constitutes a significant portion of design of a GA. While a one–one transformation between genotype and phenotype is desirable, the GA coding needs to have enough complexity to contain all possible optimal or near-optimal solutions. This translates to the problem of competing conventions in GA coding, where different chromosomes in the representation space have the same interpretation in the evaluation space. Various GAs have shown the ability to search in a complex multimodal domain and to evolve rule sets and adapt fuzzy controllers to changing operating conditions.

Numerous simulations of evolutionary fuzzy controllers have established the utility of GAs in optimizing knowledge bases for adaptive fuzzy controllers. Most of these applications differ in their approach to design an interpretation function between the phenotype and the genotype, i.e., the coding of fuzzy rule sets and membership functions, definition of fitness functions, and handling of evolutionary operations. One approach is a GA-based fuzzy controller design method that determines membership functions, number of fuzzy rules, and rule consequent parameters at the same time. Alternately, the imprecision in available information is exploited, and elements of guesswork are then employed to arrive at acceptable solutions. Still another approach is to apply GA to adaptive fuzzy control of multivariable systems and address issues such as coding of fuzzy parameters, the fitness scaling, and population size. One can evolve each fuzzy system in the form of a flexible rule base rather than as a list of parameters under a fixed set of membership functions.

Three major issues have not received the proper attention they deserve in the current literature. The first is how to utilize initial expert knowledge for a better and faster search routine. In other search engines, such as Hill Climbing, it is clear that starting from a good location can significantly improve chances for convergence to an optimal solution in a much shorter time. However, as has been shown in this book, the conventional GA applications generate a random initial population without using any expert knowledge. This, in general, will provide a more diverse population, while sacrificing convergence time. In fuzzy control applications, there is usually access to some expert knowledge which, even though it may not be the optimal solution, is often a good or reasonable solution. One can incorporate expert knowledge about limits of variables into the process of designing an interpretation function, such that impossible or obviously unreasonable phenotypes would not be evaluated. On the other hand, one can also employ heuristic rules to improve GA search (Lee and Takagi, 1993). Under this scenario, there is often a trade-off between manual knowledge and machine learning. The process of seeding the initial population with one or more experts' knowledge is also possible. The few seeded chromosomes have the chance of reproducing through mutation and crossover with other randomly generated chromosomes in the population. In this chapter, we will discuss a different method by which initial expert knowledge is incorporated into the initial population. This methodology is based on a grandparent scheme, where all individuals in the initial population are created based on a mutation from the "knowledgeable" grandparent. This scheme takes advantage of expert knowledge, while maintaining diversity needs for an effective search. An analogy to this approach is the biblical theory of creation, where all men were created from Adam and Eve.

The second issue closely follows the first. How can we utilize initial knowledge from several experts? Often, in a real-world control problem, one has access to several operators who may offer a wide array of alternative solutions. These differences in opinion come from the fact that different individuals weigh parameters differently. The real world usually involves more than one fitness parameter. For example, a system response may be evaluated for its rise time, overshoot, oscillations around steady state, and steady state error. In classical optimal control theory, a linear combination of these parameters serves as the fitness function. Changing these linear coefficients results in different solutions. In fuzzy logic and the multiexpert problem, this translates into how different individuals weigh various parameters of a system response. For example, one expert might weigh steady state error heavily, while the other might consider oscillations a more crucial factor. In this chapter, this problem is approached by using Niched GA, where information across a diverse population is preserved. In Niched GA, a population consists of several subpopulations; hence, it maintains a set of mutually supportive solutions to a given problem.

The third issue involves real-time implementation of such learning controllers. This issue remains a challenging problem due to several implementation requirements, such as the computational power, the parallel dual processing of GA and fuzzy algorithms, and the dynamic nature of adaptive fuzzy controllers. This is mainly due to the fact that a GA is strongest with a large population size. To a certain point, the larger the population size, the faster the expected rate of convergence to an optimum solution and the higher the expected rate of success. In real-time systems, though, the population consists of only one individual. Efforts in developing an on-line model of the system by which various individuals may be evaluated is also hampered by the computational requirements of such a task. This has given rise to various approaches such as incremental GA, where only one individual is evaluated in each generation.

In this chapter, the above issues are addressed in correspondence to a GA adaptive fuzzy hierarchical controller that utilizes both spatial and temporal measured data for control of distributed parameter systems. A typical example of such systems, a flexible robotic arm, is considered here. Various hardware and software components used in a real-time implementation of such an adaptive system are discussed. Membership parameters are encoded, operated upon, and optimized by GA. In order to reduce the size of the parameter set, the rule base is not encoded. Encoding rule base as well as membership functions would introduce the problem of competing conventions and redundant degrees of freedom in the optimization task.

The following section discusses the hierarchical fuzzy controller for a flexible link robot arm. As indicated, a flexible arm represents a distributed parameter system with multiple spatial and temporal optimization parameters. Then, the genetic algorithms will be discussed, the adaptive control structure and the method of incorporating initial knowledge. Also, the concepts of utilizing multiple experts in initial population and Niched GA are further explored. Next, various hardware and software implementation issues of the control structure are discussed. And finally, simulations of a flexible link robot will show improved response from the optimized controller.

7.2 Hierarchical fuzzy control for a flexible robotic link

7.2.1 A mathematical model

It is true that developing a mathematical model is not as crucial when one uses fuzzy logic for control purposes. However, it can help the control system designer by providing insight into the complex environment of the control object. Because genetic algorithms are mostly used in off-line situations, the necessity of a good mathematical model is more prudent. This section will develop a mathematical model of the flexible arm. Here, an infinite dimensional assumed mode model of a cantilevered beam is shown. An infinite number of modes in the beam's transfer function are shown to exist. Certain characteristics about the transfer function of a flexible beam are deduced.

The mode shapes of a cantilevered beam can be expressed as follows:

$$\phi_{ci} < x >= l\left(\cosh \lambda_{ci}x - \cos \lambda_{xi}x - k_{ci}\left(\sinh \lambda_{ci}x - \sin \lambda_{ci}x\right)\right) \qquad (7.1)$$

$$k_{ci} = \frac{\cos \lambda_{ci}l + \cosh \lambda_{ci}l}{\sin \lambda_{ci}l + \sinh \lambda_{ci}l} \qquad\qquad \lambda_{ci}\sqrt{\omega_{ci}\sqrt{\frac{m}{EI}}} \qquad (7.2)$$

where
k_{ci} and λ_{ci} = the roots of $1 + \cos\lambda_{ci}l \, \cosh\lambda_{ci}l = 0$
wci = the *i*th cantilevered modal frequency
l = the position along the length of the beam $0 < x < 1$
m = mass of the beam
EI = Young's modulus of elasticity

The displacement of any point on the beam can then be expressed as:

$$y<x,t>=\sum_{i=1}^{\infty}\phi_{ci}<x>q_{ci}<t>+(x+r)^{\theta}_{motor} \tag{7.3}$$

$$T_k=\frac{1}{2}\left\{\left(I_h+I_b\right)\dot{\theta}_h^2+4ml^3\dot{\theta}_h\sum_{i=1}^{\infty}\frac{\dot{q}_{ci}}{\lambda_{ci}^2}+ml^3\sum_{i=1}^{\infty}\dot{q}_{ci}^2\right\} \quad V_e=\frac{1}{2}ml^3\sum_{i=1}^{\infty}\omega_{ci}^2\dot{q}_{ci}^2 \tag{7.4}$$

Invoking the Lagrangian equation yields an infinite dimensional set of coupled second-order differential equations. As will be discussed later, as *i* increases, the significance of *i*th mode decreases. Therefore, one can closely approximate the arm behavior by retaining the first *n* elastic modes as follows:

$$M_c\ddot{q}+K_cq=Q \tag{7.5}$$

$$M_c=\begin{bmatrix} I_T & \dfrac{2ml^{-2}}{\lambda_{cl}^2} & \cdots & \dfrac{2ml^3}{\lambda_{cn}^2} \\ \dfrac{2ml^3}{\lambda_{c1}^2} & ml^3 & 0... & 0 \\ . & & . & \\ \dfrac{2ml^3}{\lambda_{cn}^2} & 0 & ...0 & ml^3 \end{bmatrix} \qquad q=\begin{bmatrix} \theta_{motor} \\ q_{ci} \\ . \\ q_{ci} \\ . \\ q_{cn} \end{bmatrix} \tag{7.6}$$

$$K_c=\begin{bmatrix} 0 & . & . & . & . & 0 \\ 0 & ml^3\omega_{c1}^2 & 0 & . & . & 0 \\ . & 0 & . & 0 & . & 0 \\ . & . & 0 & \omega_{cl}^2 & 0 & 0 \\ . & . & . & 0 & . & 0 \\ 0 & . & . & . & 0 & ml^3\omega_{c1}^2 \end{bmatrix} \qquad Q=\begin{bmatrix} T \\ 0 \\ . \\ . \\ . \\ 0 \end{bmatrix} \tag{7.7}$$

This can be arranged in the following form, or can be transformed to a state space form,

$$\ddot{q}=M_c^{-1}\left(Q-K_cq\right) \tag{7.8}$$

where

$$\begin{bmatrix} \dot{q} \\ \ddot{q} \end{bmatrix} = \begin{bmatrix} 0 & I \\ -M_c^{-1} & 0 \end{bmatrix} \begin{bmatrix} q \\ \dot{q} \end{bmatrix} + \begin{bmatrix} 0 \\ M_c^{-1}I \end{bmatrix} T \qquad (7.9)$$

where $I = [1,0,\ldots,0]^T$.

Asymptotically stabilizing a transfer function with an infinite number of poles on the $j(\omega)$-axis is a difficult control task. However, as is shown by Kotnick et al. (1988), most of the poles of a physical system are slightly damped, i.e., they are all but one inside the left half plane. Second, as the poles move away from the origin, their real part becomes more negative. This implies that they die out faster. Third, as the poles move away from the origin, they contribute less to the overall system's transfer function. Therefore, a reduced-order model can approximate the full-order model closely. In this chapter, in addition to the rigid mode, three of the flexible modes are accounted for, $\omega = [7, 23, 62]$.

7.2.2   Separation of spatial and temporal parameters

From the previous section, the difference between a distributed parameter model and a lumped parameter model is exhibited through the mode shape function. A lumped parameter system has certain desirable temporal characteristics, such as fast rise time, small overshoot, and small oscillations at steady state. A distributed parameter system, such as a flexible robot, involves additional desirable spatial characteristics such as low spatial oscillations along the length of the flexible link during movement and after achieving final value.

The additional spatial consideration has prompted the following bilevel hierarchical control architecture as is shown in Figure 7.1. The second level of hierarchy monitors the behavior of the distributed-parameter system (robot arm) and extracts spatial features such as Straight, Oscillatory, and Gently Curved. This is accomplished by using a fuzzy feature extraction technique, which incorporates line curvature data computed over a finite number of strain sensors at discrete points along the length of the robot arm. The extracted fuzzy features such as Straight, Gently Curved, together with temporal measurements, such as the angular position and velocity measurements of the tip of the robot arm, constitute a basis of information for the first level of the hierarchical fuzzy controller. The first level of hierarchical controller determines the torque to be applied at the arm's base.

7.2.3   The second level of hierarchical controller

Development of the lower level controller depends on a proper choice of features. Here, the average and variance of the strain gauge measurements are calculated and used to classify and extract from various arm motions.

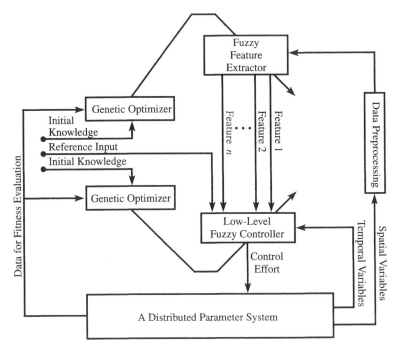

Figure 7.1 GA-based learning hierarchical control architecture.

7.2.3.1   Line-curvature analysis

Consider a hypothesized dynamic behavior of curvature $\delta(k)$ versus position k. Figure 7.2 shows this feature. Let us define the curvature mean as follows:

$$\bar{\delta} = \sum_k \frac{\delta(k)}{n} \tag{7.10}$$

where n is number of total points (sensors) distributed evenly throughout the arm's length. Similarly, we define the curvature as follows:

$$\sigma_\delta = \sum_k \frac{\left(\delta(k) - \bar{\delta}\right)^2}{n} \tag{7.11}$$

7.2.3.2   The rule base

Figure 7.1 shows the closed-loop hierarchical controller for the flexible arm. The second level of hierarchy uses measured data from the arm to determine the features of the arm. One can now describe a mapping between characteristics of the line curvature to features of the line. This mapping constitutes the basis of the expert knowledge from which the fuzzy rules are drawn.

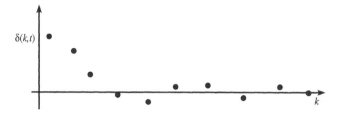

Figure 7.2 Curvatures at a time instant t.

Table 7.1	Features Extracted from a Plot of Incremental Curvature		
$\delta(k)$ versus k	Shape Features	δ	σ_δ
Positive horizontal line	Gently curved	Positive	Zero
Negative horizontal line	Gently curved	Negative	Zero
Zero horizontal line	Straight	Zero	Zero
Oscillatory line	Oscillatory	Irrelevant	Nonzero

Table 7.1 lists three fuzzy features of the arm and their correspondence to fuzzy sets of $\bar{\delta}$ and σ_δ. Therefore, a fuzzy rule may be constructed as below:

If $\bar{\delta}$ is zero and σ_δ is zero
Then Straight is high, Oscillatory is low, and Gently Curved is zero.

7.2.4   The lower level of hierarchy

The objective of the fuzzy controller is to track a desired trajectory while keeping oscillations along the robot arm to a minimum. First, a table of expert knowledge is presented, and based upon this table, certain rules are formulated. The genetic algorithm will upgrade this knowledge base, based on the performance fitness of the rule base.

Visualizing different arm features is an important part of controller design. Figure 7.3 illustrates four primary combinations of arm features. Control depends, to a large extent, on the combination of these features. Velocity$_{\text{Error}}$ and Position$_{\text{Error}}$ have nine possible combinations. Therefore, a total of 36 (4×9) rules completely span the state space. A sample rule in the proposed fuzzy controller may be as follows:

If Straight is high, Oscillatory is low, Error$_\theta$ is zero, and Velocity is positive
Then Torque is negative

For a complete list of fuzzy rules for the flexible link's curvature classification, the reader can refer to the work of Akbarzadeh-T (1998).

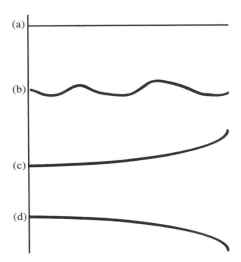

Figure 7.3 Primary shape features of the arm: (a) *Straight*, (b) *Straight and Oscillatory*, (c) *Negative Gently Curved*, and (d) *Positive Gently Curved*.

7.3   Genetic algorithms in knowledge enhancement

Application of genetic algorithms (GA) to knowledge enhancement for fuzzy controllers involves several aspects. The first aspect is how to code a string to represent all the necessary degrees of freedom for search in the fuzzy knowledge domain. While a one-to-one transformation between genotype and phenotype is desirable, the GA coding needs to have enough complexity to contain all possible optimal or near-optimal solutions. This translates to the problem of competing conventions in GA coding, where different chromosomes in the representation space have the same interpretation in the evaluation space. The second aspect is how to incorporate existing expert knowledge into the GA optimizing algorithm, and, generally, how to take advantage of several experts' opinions in creation of initial population. Conventional applications of GA fuzzy applications suggest a random initial population. However, it is intuitively clear that any search routine could converge faster if starting points are good solutions.

7.3.1   Interpretation function

An interpretation function is the transformation function between the representation (genotype) space and the evaluation (phenotype) space. This is perhaps the most important stage of GA design, and it significantly affects the algorithm's performance. Two important general categories of expert knowledge are considered below:

- *Membership function*: General shape (triangular, trapezoidal, sigmoidal, Gaussian, etc., see Appendix A)

- *Rule base*: Fuzzy associate memory, disjunctive (OR) and conjunctive (AND) operations among antecedents in the rule base

If both the rule base and the membership functions are coded together in a representation space, the problem of competing conventions arises, and the landscape would unnecessarily become multimodal. In other words, there would be different chromosomes representing the same nonlinear function (the control surface) in the evaluation space. In this representation, even though the GA might come up with two fit individuals (with two competing conventions), the genetic operator, such as crossover, will not yield fitter individuals.

In this chapter, without any loss of generality, only triangular membership functions are coded for optimization. Figure 7.4 illustrates a triangular membership function, where its three determining variables (a,b,c) are shown. Assuming a normalized membership function, the three variables are real numbers between 1 and 1. The coding in GA is performed as follows, with the real variable *a* first mapped into an n-bit signed binary string, where the highest bit represents the sign (Figure 7.5). This way, the variable *a* can take on 2^n different values. Then, the binary number is aggregated with other rabbit binary numbers to construct the phenotype representation.

7.3.2   *Incorporating initial knowledge from one expert*

The method presented here is based on a grandparent scheme, where the grandparent is the genotype representation of one expert's rule set and membership functions. All members of the initial population are binary

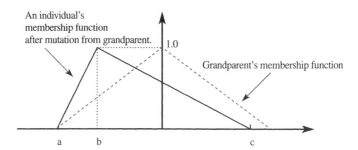

Figure 7.4 Triangular membership function with three variables.

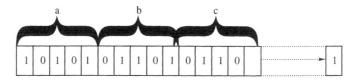

Figure 7.5 Chromosome coding in representation space.

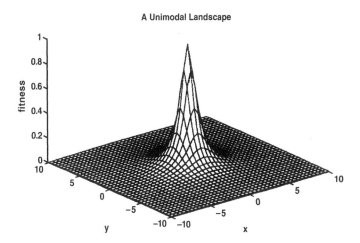

Figure 7.6 A standard GA may be sufficient to exploit the landscape with one expert knowledge.

mutations of the grandparent. The mutation rate is a significant factor, because as the mutation rate increases, diversity among members of the initial population increases. Figure 7.6 illustrates such a unimodal landscape.

One of the possible concerns about this approach is the diversity issue. If all members of the initial population are to be derived from one individual (the grandparent), will there be enough diversity among the initial knowledge altogether and start with a totally random population. Let's look at two extreme situations.

- *First extreme*: The mutation rate is set to 0. Hence all members of population are exactly alike. The initial population will consequently have a high average fitness and very low diversity. It is difficult for this initial population to recombine to create a more fit individual unless the mutation rate is set to a high value. This is because all members are alike, and therefore, crossover is not a productive operator.
- *Second extreme*: The mutation rate is set to 1. In this case, we are essentially back to a totally random population with a low average fitness and a high diversity. This type of initial population will, in general, result in a faster learning rate; but since the initial fitness is low, it will take a long time to converge.

From the above, it is clear that in most cases, the mutation rate should be set to a value between 0 and 1. A lower mutation rate indicates a higher degree of confidence that the optimal gene is in close proximity of the grandparent's, the expert, gene. A higher mutation rate indicates a low degree of confidence in the expert and the need for exploring the rest of the representation space more fully and with a higher diversity. In short, the design adds a control variable, the mutation rate, as a new parameter by which the GA designer can weigh diversity versus average fitness of the initial population.

7.3.3   Incorporating initial knowledge from several experts

The evolution in nature is not limited to manipulating genes and chromosomes. In fact, there are GA-based models of thought patterns in the brain which continuously evolved through reproduction, mutation, and recombination (crossover). This explains the diversity in human minds, where each human differs in perspective and opinion if faced with the same constraints and criteria. It is therefore of no surprise that if several experts are interviewed in regards to a control system, they would each give differing heuristics.

This difference in heuristics suggests multimodality in the knowledge domain. Consequently, it is important to use a technique such as GA to properly exploit the multimodal nature of the problem. For this purpose, Niched GA is used, which allows evolution of several subpopulations. Figure 7.7 illustrates a multimodal landscape where Niched GA would be able to simultaneously exploit several optimal points. In this representation, the whole population is analogous to the people of the world. And, the subpopulations are analogous to the people of the same country. As is the case with our world, the Niched GA is designed such that most of the recombination takes place among members of the same subpopulation. Occasionally, there is reproduction across subpopulations, which allows exploitation of new spaces.

7.4   Implementation issues

7.4.1   Software aspects

During the past two decades, many commercial packages and programs have been developed for analysis and synthesis of fuzzy control systems (Jamshidi

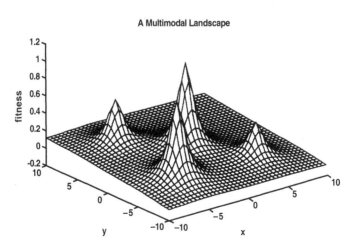

Figure 7.7 This landscape requires Niched GA to explore all optimal solutions.

et. al., 1993) Some of the more prominent packages are FULDEK — Fuzzy Logic Development Kit (Jamshidi, 1996) and MATLAB's® Fuzzy Logic Tool Box (Zilouchian and Jamshidi, 2001) as well as numerous software programs for neural networks and genetic algorithms (see also Chapter 9).

The software described in this chapter, a dynamic software program, called SoftLab (Akbarzadeh-T, 1998),was used in this work, which allowed for adaptation parameters to be passed back and forth between an optimization algorithm and the rule-based controller through a block of memory. In our particular hardware arrangement, this memory block is located on a Digital Signal Processor (DSP) board, where the Dynamic Fuzzy runs in real time. The optimizing algorithm can be arbitrary as long as communication is performed through the memory block. Therefore, another fuzzy in module or a module based on GA or neural networks, running on the PC, can be used to modify a controller's rule set. SoftLab is a dynamic program that allows for incorporating new rules in the rule set while disposing of undesirable rules, changing of the number of rules, and tuning the input and output membership functions. The combination of DSP board and the Pentium's main CPU allows for simultaneous dual processing of tasks. The GA optimizing algorithm runs on the Pentium, and the fuzzy controller runs on the DSP board. Because the tuning of the fuzzy controller is performed by adjusting only a memory block, the update time during each adaptation becomes minimal. In other words, the whole fuzzy routine no longer needs to be updated and compiled after each adaptation. The main routine can permanently rest in the DSP board's memory, and only a small memory block must be updated each time with the new parameters.

7.4.2   Hardware aspects

DSP Research's TIGER 30 board (Akbarzadeh-T, 1998) utilizes Texas Instruments TMS32OC30 DSP chip20 and is fully compatible with a Pentium computer processor board. The TMS32OC30 digital signal processor has a powerful instruction set, operates at 40 MFLOPS, and can process data in parallel. The TIGER board has the ability of being interfaced with a PC, therefore, allowing dual processing of the genetic algorithm as well as the fuzzy logic controller. The fuzzy logic controller is continuously processed in the DSP board at 1 kHz. Parameters that identify the rule base (membership functions and fuzzy associative memory) are stored in a block of memory residing on the DSP board. This block of memory is modified by the PC that runs the code for genetic algorithms. Each time the PC has completed an optimization task, a signal is sent to the DSP board. The DSP board then modifies its parameters accordingly and sends data on the performance of the fuzzy controller back to the PC. Figure 7.8 shows a block diagram of the hardware setup. A generic data acquisition system was designed to serve as a medium for data transfer between the control system and the DSP board. The data acquisition system is capable of asynchronous and 16-bit data transfer. A program written in C is used to transfer data to and from the control system.

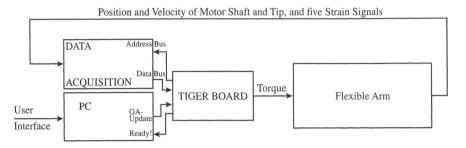

Figure 7.8 A block diagram of the hardware setup.

7.5 *Simulation*

In this work, simulation is used to predict the performance of the real-time integrated system that will be realized in the near future. In order to minimize the GA parameter set, GA is only applied to optimize parameters related to the input membership parameters of the higher level of hierarchy, as is shown in Figure 7.1. Other parameters in the knowledge base are not allowed to vary. This will reduce simulation in time and will still demonstrate the potential utility of GA. The following fitness function was used to evaluate various individuals within a population of potential solutions:

$$fitness = \int_{t_i}^{t_f} \frac{1}{e^2 + x^2 + 1} dt \qquad (7.12)$$

where e represents the error in angular position, and x represents overshoot. The above fitness function was chosen to contain important parameters in a system's time response: rise time, steady state error, oscillations, and overshoot. The fitness function is inversely proportional to the error and overshoot in the system, and consequently, a fitter individual with a lower overshoot and a lower overshoot error (shorter rise time) in its time response.

Figure 7.9(a) shows the time response of the GA optimized fuzzy controller compared to that of a non-GA fuzzy controller. A constant population size of 40 individuals was used. After only ten generations, the rise time is improved by 0.34 sec (an 11% improvement), and the overshoot is reduced by 0.07 radians (a 54% improvement). Figure 7.9(b) shows the average fitness of each generation. The mutation rate for generating the initial population was set at 0.3. The resulting initial population has an average fitness of 15.8, which converges to 28 after only 10 generations. The mutation rate for creating the following generations was set at 0.033. The probability of crossover was set to 0.6.

A simulation of the hierarchical controller was also performed with a mutation rate of 1, i.e., a random population. The fitness value remained

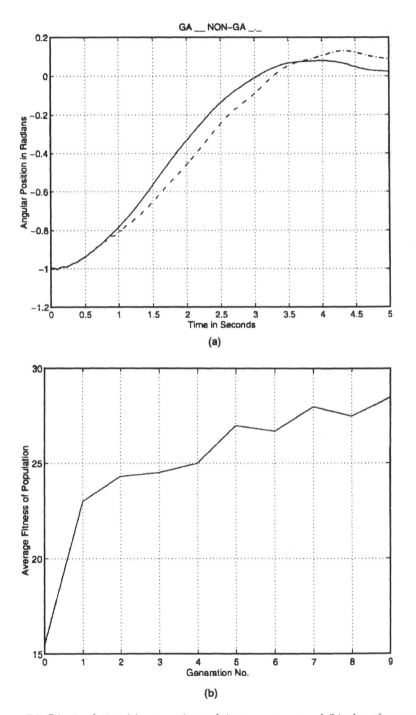

Figure 7.9 GA simulation (a) comparison of time responses and (b) plot of average fitness.

close to zero for the first ten generations, and the system response was divergent. Therefore, the results are not illustrated here. This shows the difficulty of the optimization task for flexible robot control and, consequently, the necessity of incorporating initial knowledge.

7.6 Conclusions

In this chapter, we touched on several aspects of applying the concept of evolutionary optimization to fuzzy logic knowledge development. We first discussed the process of designing the interpretation j function with consideration for importance of avoiding competing convention. The fuzzy knowledge base is divided in two major categories: the membership functions and the rule base. A novel method for integrating initial knowledge from an expert in GA search was then discussed. The proposed method allows the mutation parameter to serve as a degree of confidence. The more confidence we have in our initial knowledge, meaning that the optimal value is in close proximity in the search landscape, the lower the rate of mutation. If we have no confidence in our expert's knowledge, then the mutation rate is set to 1 to create a totally random population with a high diversity.

The above process is then applied to a flexible robot hierarchical controller with two levels. Flexible robot operation in unstructured environments requires autonomous adaptability to the environmental conditions. When using fuzzy logic controllers, GA can offer a suitable approach to achieving adaptation. It is shown that the problem of flexible robot control is a difficult problem using conventional GA search techniques. By using the presented method of incorporating initial knowledge, the average fitness of the initial population is improved to 15.8. And, the algorithm converged within 10 generations.

In this chapter, we also outlined various aspects of implementing such an adaptive fuzzy controller. The processing of the genetic algorithm and fuzzy controller is often too demanding on a single processor. Therefore, the hardware proposed here offers a versatile quick data acquisition system combined with a fast Digital Signal Processing board and a Pentium processor to accommodate the dual processing requirement. In addition, a Dynamic Fuzzy software package is developed which is compatible with the GA and allows for changing the knowledge base in real time. Parameters in the knowledge base can be modified in real time using SoftLab (Akbarzadeh-T, 1998). Simulation of a flexible link robot shows that GA optimization is capable of significantly improving response. Future directions of this research involve experimental testing of the improved fuzzy controller and finding alternate methods to approach genetic algorithms in real time.

References

Akbarzadeh-T, M.-R., Ph.D. *Fuzzy Control and Evolutionary Optimization of Complex Systems*, Dissertation, Department of EECE and the ACE Center, University of New Mexico, 1998.

Akbarzadeh-T, M.-R. and Jamshidi, M., Evolutionary fuzzy control of a flexible link, *Intelligent Automation and Soft Computing*, 3, 1, 77–88, 1997.

Jamshidi, M., *Large-Scale Systems — Modeling, Control and Fuzzy Logic*, Vol. 8, Series on EIMS, Jamshidi, M., Series Ed., Prentice Hall, New York, 1996.

Jamshidi, M., Vadiee, N., and Ross, T.J., *Fuzzy Logic and Control with Software and Hardware Applications*, Vol. 3, Series on EIMS, Jamshidi, M., Series Ed., Prentice Hall, New York, 1993.

Kotnik, P.T., Yurkovich, S., and Ozguner, U., Acceleration feedback for control of a flexible manipulator arm, *Journal of Robotic Systems*, 5(3), 181–196, 1988.

Lee, M.A. and Takagi, H., Integrating design stages of fuzzy systems using genetic algorithms, *Proceedings of the 1993 IEEE International Conference on Fuzzy Systems*, San Francisco, California, pp. 612–617, 1993.

Zilouchian, A. and Jamshidi, M., *Intelligent Control Systems with Soft Computing Methodologies*, CRC Press, Boca Raton, Florida, 2001.

chapter eight

Autonomous robot navigation through fuzzy-genetic programming*

8.1 Introduction

Real-time intelligent robot controllers are required to achieve the level of autonomy necessary in unstructured or "nonengineered" operating domains. For mobile robots, the operating domain, whether an indoor industrial workplace, outdoor paved road, or planetary surface, dictates acceptable design and control approaches. One common requirement, independent of the operating domain, is a capability for intelligent navigation. Traditional approaches based on sequential task decomposition have met with difficulty in achieving real-time response. Controllers are now being developed that are modeled after natural processes and that advocate a behavioral decomposition of tasks with quasi-parallel execution. These behavior-based systems facilitate real-time intelligence by decomposing motion control capabilities into a set of special-purpose routines that operate concurrently.

Robust behavior in autonomous robots requires that uncertainty be accommodated by the robot control system. Fuzzy logic is particularly well suited for implementing such controllers due to its capabilities of inference and approximate reasoning under uncertainty. As such, we have focused our research efforts on developing an architecture that incorporates fuzzy control theory into the framework of behavior-based control for mobile robots. A mobile robot's behavior is encoded as a fuzzy rule base that maps relevant sensor inputs into control outputs according to a desired control policy. Many fuzzy controllers proposed in the literature utilize a monolithic rule-base structure. That is, the precepts that govern desired system behavior are encapsulated as a single collection of *if–then* rules and, in most instances,

* This work is based on Tunstel, Jr., E., Lippincott, T., and Jamshidi, M., *International Journal on Intelligent Automation and Soft Computing*, 3, 1, 37–50, 1997. NASA Grant numbers NAG 2-1196 and NAG 2-1480 supported this work.

the rule base is designed to carry out a single control policy or goal. In order to achieve autonomy, mobile robots must be capable of achieving multiple goals with priorities that may change with time. Thus, controllers should be designed to realize a number of task-achieving behaviors that can be integrated to achieve different control objectives. This requires formulation of a large and complex set of fuzzy rules. In this situation, a potential limitation to the utility of the monolithic fuzzy controller becomes apparent. Because the size of complete monolithic rule bases increases exponentially with the number of input variables (Raju et al., 1991), multi-input systems can potentially suffer degradations in real-time response. This is a critical issue for mobile robots operating in dynamic surroundings. *Hierarchical* rule structures can be employed to overcome this limitation, such that the size of the rule base increases linearly with the number of input variables (Raju et al., 1991; Bruinzeel et al., 1995).

This chapter describes a behavior-based robot navigation control system. Autonomy is achieved within a hierarchical structure of fuzzy behaviors in which low-level navigation behaviors are realized as fuzzy logic controllers, while higher-level coordination behaviors are implemented as fuzzy decision systems. This hierarchy of fuzzy rule bases enables distribution of intelligence among special-purpose *fuzzy behaviors*. Its structure is motivated by the hierarchical nature of behavior as hypothesized in ethological models of animal behavior. A fuzzy coordination scheme is described that employs weighted decision making based on contextual behavior activation. We also address the problem of automatic discovery and learning of coordination rules using genetic programming (GP) — an approach to computational intelligence. Performance results are presented on coordination behaviors evolved for the goal-seeking portion of a mobile robot behavior hierarchy. Simulated navigation experiments highlight interesting aspects of the decision-making process arising from behavior interaction (Tunstel et al., 1997). The focus here is on indoor navigation; however, the architecture and approach can be extended for implementation on autonomous rovers or other vehicles designed for outdoor traversal in natural terrain.

8.2   Hierarchical fuzzy-behavior control

The behavior control paradigm has grown out of an amalgamation of ideas from ethology, control theory, and artificial intelligence (McFarland, 1971; Brooks, 1986). High-level motion control is decomposed into a set of special-purpose behaviors that achieve distinct tasks when subject to particular stimuli. Clever coordination of individual behaviors results in emergence of more intelligent behavior suitable for dealing with complex situations. The paradigm was initially proposed by Brooks (1986) and was realized as the "subsumption architecture," wherein a behavior system is implemented as distributed finite state automata. Until recently (Pin and Watanabe, 1994; Tunstel and Jamshidi, 1994 and 1997; Jamshidi, 2001), most behavior controllers have been based on crisp (nonfuzzy) data processing and binary logic-based reasoning. In contrast to their crisp counterparts, fuzzy behaviors are synthesized as fuzzy rule bases, i.e., collections of a finite set of fuzzy *if–then* rules. Each behavior is encoded with a distinct control policy governed by fuzzy inference. Thus, each fuzzy behavior is similar to the conventional fuzzy controller in that it performs an inference mapping from some input space into some output space. If X and Y are input and output universes of discourse of a behavior with a rule base of size n, the usual fuzzy if–then rule takes the following form:

$$\text{IF } x \text{ is } A_1 \text{ THEN } y \text{ is } B_j \tag{8.1}$$

where x and y represent input and output fuzzy linguistic variables, respectively, and A_1 and B_i ($B_i = 1,...,n$) are fuzzy subsets representing linguistic values of x and y. In the mobile robot controller, the input x refers to sensory data or goal information, and y refers to motor control signals or behavior activation levels. The antecedent consisting of the proposition "x is A_1" could be replaced by a conjunction of similar propositions; the same holds for the consequent "y is B_i".

8.2.1   Behavior hierarchy

The proposed architecture is a conceptual model of an intelligent behavior system and its behavioral relationships. Overall robot behavior is decomposed into a bottom-up hierarchy of increased behavioral complexity, in which activity at a given level is a function of behaviors at the level(s) below. A collection of *primitive behaviors* resides at the lowest level, which we refer to as the primitive level. These are simple, self-contained behaviors that serve a single purpose by operating in a reactive (nondeliberative) or reflexive (memoryless) fashion. They perform nonlinear mappings from different subsets of the robot's sensor suite to common actuators. Each primitive behavior exists in a state of solipsism (a primitive behavior is only cognizant of its own purpose and knows nothing about other behaviors), and alone, would be insufficient for autonomous navigation tasks. Primitive behaviors are building blocks for more intelligent *composite behaviors*. Their capabilities can be modulated through synergistic coordination to produce composite behavior(s) suitable for goal-directed operations.

Mobile robots operating in nonengineered domains must be capable of reliable navigation in the presence of static and dynamic obstacles (e.g., humans and other moving robots). It is preferred that such robots be designed to autonomously navigate, in an equally effective manner, in both sparsely populated environments (e.g., an overnight security robot patrolling an office building) and in cluttered environments (e.g., a robot transporting material on a busy factory floor). Our approach to providing the necessary autonomy for such robots is to construct behavior hierarchies suitable for the task(s) and operating environment. A behavior hierarchy for indoor

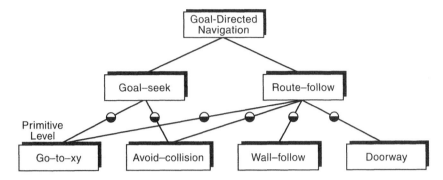

Figure 8.1 Hierarchical decomposition of mobile robot behavior.

navigation might be organized as in Figure 8.1. It implies that goal-directed navigation can be decomposed as a behavioral function of *goal-seek* (collision-free navigation to some location) and *route-follow* (assuming some direction is given in the form of waypoints or a path plan). These behaviors can be further decomposed into the primitive behaviors shown, with dependencies indicated by the adjoining lines. *Avoid-collision* and *wall-follow* are self-explanatory. The *doorway* behavior guides a robot through narrow passageways in walls; *go-to-xy* directs motion along a straight-line trajectory to a particular location. The circles represent weights and activation thresholds of associated primitive behaviors. As described below, fluctuations in these weights are at the root of the intelligent coordination of primitive behaviors, which leads to adaptive system behavior. The hierarchy facilitates decomposition of complex problems as well as run-time efficiency by avoiding the need to evaluate rules from behaviors that do not apply.

As an additional example, consider the outdoor navigation problem encountered by natural terrain vehicles such as planetary rovers. Autonomous rovers must be capable of point-to-point navigation in the presence of varying obstacle (rocks, boulders, dense vegetation, etc.) distributions, surface characteristics, and hazards. Often, the task is facilitated by knowledge of a series of waypoints, furnished by humans, which lead to designated goals. In some cases, such as exploration of the surface of Mars (Matthias et al., 1995; Matijevic and Shirley, 1996), this supervised autonomous control must be achieved without the luxury of continuous remote communication between the mission base station and rover. Time delays between Earth and Mars can be anywhere between 6 and 41 min. Considering these and other constraints associated with rover navigation, suitable behavior hierarchies similar to the hypothetical one shown in Figure 8.2 could be constructed. In this figure, the behavioral functions of *goal-seek, route-follow,* and an additional composite behavior, *localize,* are decomposed into a slightly different suite of primitive behaviors. The design of behaviors at the primitive level would be tailored to the navigation task and an environment with characteristics of natural terrain.

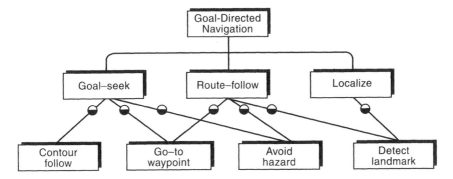

Figure 8.2 Hypothetical behavior hierarchy for rover navigation.

Note that decomposition of behavior for a given mobile robot system is not unique. Consequently, suitable behavior repertoires and associated hierarchical arrangements are arrived at following a subjective analysis of the system and the task environment. The total number, and individual purpose, of fuzzy behaviors in a given behavior hierarchy are indicative of the problem complexity and can be conveniently modified as required.

8.3   Coordination by behavior modulation

During any sufficiently complex navigation task, interactions in the form of behavioral cooperation or competition occur when more than one primitive behavior is active. These forms of behavior are not perfectly distinct; they are extremes along a continuum (Staddon, 1983). Instances of behavior throughout the continuum can be realized using *behavior modulation*, which we define as the autonomous act of regulating, adjusting, or adapting the activation level of a behavior to the proper degree in response to a context, situation, or state perceived by an autonomous agent. In a given implementation, the "proper degree" is governed by the desired behavioral response of the agent. In our hierarchical approach, coordination is achieved by weighted decision making and behavior modulation embodied in a concept called the *degree of applicability* (DOA), which is a measure of the instantaneous level of activation of a behavior. It can be thought of in ethological terms as a motivational tendency of the behavior. Fuzzy rules for coordination are formulated to include weighting consequents, which modulate the DOAs of lower-level primitives. The DOA, $q \in [0,1]$, of primitive behavior j is specified in the consequent of *applicability rules* of the following form:

$$\text{IF } x \text{ is } A_j \text{ THEN } \alpha_j \text{ is } D_i \qquad (8.2)$$

Where, as in Equation (8.1), x is an input linguistic variable and A_1 is a fuzzy subset representing its linguistic value. D_i is a fuzzy subset specifying the linguistic value (e.g., " *high* ") of *a* for the situation prevailing during the current control cycle. This feature allows certain robot behaviors to influence the overall behavior to a greater or lesser degree depending on the current situation. It serves as a form of motivational adaptation, because it causes the control policy to dynamically change in response to goals, sensory input, and internal state. Thus, composite behavior is meta-rule-bases that provide a form of the ethological concepts of inhibition and dominance. Behaviors with maximal applicability ($\alpha_{max <=1}$) can be said to dominate, while behaviors with partial applicability ($0 < \alpha < \alpha_{max}$) are said to be inhibited. These mechanisms allow exhibition of behavioral responses throughout the continuum. This is in contrast to crisp behavior selection that typically employs fixed priorities which allow only one activity to influence the robot's behavior during a given control cycle. The coordination scheme includes behavior selection as a special case when the DOA of a primitive behavior is nonzero and above its activation threshold, while others are zero or below threshold. When this occurs, the total number of rules to be consulted on a given control cycle is reduced. The reduction in rule evaluations is not as dramatic or static as in the strict rule hierarchies (Nordin and Banshaf, 1995), because we are dealing with a *behavior* hierarchy that achieves interacting goals. As such, the number of rules consulted during each control cycle varies dynamically as governed by the DOAs and thresholds of the behaviors involved.

Coordination and conflict resolution are achieved within the framework of fuzzy logic theory via operations on fuzzy sets (Tunstel, 1995). Fuzzy rules of each applicable primitive behavior are processed, yielding respective output fuzzy sets. These fuzzy sets are equivalent to the result produced by rule-base evaluation in conventional fuzzy controllers *before* applying the defuzzification operator. Following consultation of applicable behaviors, each fuzzy-behavior output is weighted (multiplied) by its corresponding DOA, thus effecting its activation to the level prescribed by the composite behavior. The resulting fuzzy sets are then aggregated using an appropriate t-conform operator (i.e., a fuzzy union/disjunction operator such as arithmetic maximum) and defuzzified to yield a crisp output that is representative of the intended coordination. Because control recommendations from each applicable behavior are considered in the final decision, the resultant control action can be thought of as a consensus of recommendations offered by multiple experts.

8.3.1   Related work

This strategy for multiple behavior coordination was developed to enable robust autonomous performance. It represents an approach that is particularly suitable in the context of fuzzy-behavior hierarchies. Several instances of independent research have converged to similar ways of approaching autonomous mobile robot navigation (Saffioni, et al., 1995; Moreno et al., 1993; Michaud, et al., 1996; Correia and Steiger-Gargao, 1995).

Experience thus far has revealed that in many practical instances, fuzzy control alone is insufficient for addressing complex intelligent control problems of robotics. Furthermore, humans often find it difficult to design knowledge-based control systems with interacting rule bases, particularly in the absence of experts or sufficient knowledge of the problem. In what follows, we address the problem of automatic discovery and learning of coordination rules for use in the proposed coordination strategy. This problem has been previously approached in the contexts of other coordination schemes by using reinforcement learning (Bonarini, 1994) and hybrids of reinforcement and neural networks (Beom et al., 1994; Gachet et al., 1994). In Tunstel and Jamshidi (1996, 1997), the potential of the GP paradigm for learning fuzzy rule bases for low-level regulation and tracking types of problems was demonstrated. The reader interested in details of its implementation may consult the reference. Using this earlier work as a foundation, we apply our GP approach here to higher-level behavior coordination.

8.4 Genetic programming of fuzzy behaviors

The GP paradigm (Koza, 1992) computationally simulates the Darwinian evolution process by applying fitness-based selection and genetic operators to a population of parse trees representing computer programs of a given programming language. It departs from the conventional genetic algorithm primarily with regard to its representation scheme. Structures undergoing adaptation are executable hierarchical programs of dynamically varying size and structure, rather than numerical strings. In our GP system, a population comprised of fuzzy rule bases that are candidate solutions to the problem evolves in response to selective pressure induced by their relative fitnesses for implementing the desired behavior.

8.4.1 Rule discovery

In the process of learning fuzzy rules, GP manipulates the linguistic variables directly associated with the fuzzy behaviors. The dynamic variability of the representation allows for rule bases of various sizes, thus enhancing population diversity, which is important for the success of the GP system. It also increases the potential for discovering rule bases of smaller sizes than necessary for completeness (a rule base is complete if there exists a rule for any valid combination of inputs) but sufficient for realizing desired behaviors. In this work, lower and upper bounds of 8 and 27 were imposed on rulebase size. The search space is contained in the set of all possible rule bases that can be composed recursively from a set of *functions* and a set of *terminals*. The function set consists of components of the generic fuzzy *if–then* rule and common fuzzy logic connectives, i.e., functions for antecedents, consequents, fuzzy intersection, rule inference, and fuzzy union. Each rule base is an executable program that evaluates to an output fuzzy set resulting from fuzzy inference. The terminal set is made up of the input and output linguistic variables and prespecified membership functions associated with the desired behavior.

Given suitable function and terminal sets, GP proceeds by randomly generating an initial population of behaviors. This is followed by evaluating each behavior in the current population and applying, primarily, reproduction and crossover to behaviors selected with a probability based on fitness. Crossover is done by swapping portions of parse trees from two parent behaviors such that proper syntax is preserved. Upon satisfaction of termination criteria, the GP result is the best-fit behavior that appeared in any generation. In addition to this generational process, a steady-state evolution can be applied as in the Steady State Genetic Algorithm (SSGA) (Syswerda, 1991), which has recently been applied in GP for behavior evolution (Reynolds, 1994; Nordin and Banshaf, 1995). In the Steady State Genetic Programming (SSGP) approach, the concept of "generations" does not exist. Instead, on each iteration following creation of the initial population, only two new offspring are produced; the offspring replace two of the worst individuals in the population. In our approach, parent behaviors are selected to produce offspring by tournament, while behaviors to be removed are chosen randomly from the set of below-average behaviors in the current population.

8.5 Evolution of coordination

When behaviors compete for control of the robot by recommending different control actions for common actuators, the problem becomes one of coordinating multiple task-achieving behaviors and resolving conflicts among them. In order to formulate suitable coordination rules, one must first decide what the DOAs of low-level behaviors should be in all practical situations perceived from sensory input. Formulation of such rules for coordination is not entirely intuitive, and expert knowledge about how to concurrently coordinate primitive behaviors is not readily available. We apply GP for this purpose. A fuzzy coordination behavior for goal seeking, comprised of rules in the form of Equation (8.2), is what we wish to evolve. In the current implementation, applicability rules used by *goal-seek* to modulate the underlying primitive behaviors consider three instantaneous input states — the range to the nearest obstacle, the distance from the goal, and the angular heading to the goal. The consequents of these rules prescribe a DOA for each primitive behavior.

8.5.1 Behavior fitness evaluation

During evolution, each behavior in the current population is evaluated via simulation in a number of indoor *fitness cases* subject to an upper time limit of 200 sec. In this work, $n_f = 5$ fitness are used; the simplest and most difficult of these are illustrated in Figure 8.3(a) and (b), respectively. An X indicates goal locations in the figure, the robot is depicted as an octagonal icon with a radial line designating its initial heading, and its range sensor horizon is indicated by the shaded regions of Figure 8.3(a). In each case, the dimension of the indoor space is 10 m × 10 m. Each fitness case was chosen to represent situations likely to be encountered in indoor environments.

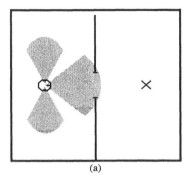 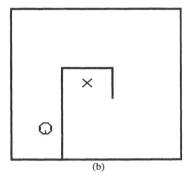

(a)   (b)

Figure 8.3 Example fitness cases.

For a given behavior, the score of a trial run through fitness case i is given by the following:

$$Si = \begin{cases} 100 \\ \dfrac{100}{\gamma(1+10eN)} \end{cases} \tag{8.3}$$

where e_N is the normalized residual distance to the goal in the case of a time-out or collision. The parameter $\gamma = 2$ if a collision occurs; otherwise, $\gamma = 1$. That is, the score for an unsafe trial is half of that for a collision-free trial with all else being equal (see Figure 8.4). The overall fitness of the behavior is the average score over all nf fitness cases:

$$F = \frac{1}{nf} \sum_{i=1}^{nf} Si \tag{8.4}$$

Thus, the highest possible score, and hence fitness, is 100. In evolutionary algorithms, such as GP, it is important that the fitness function map observable parameters of the problem into a spectrum of values that differentiate the performance of individuals in the population. If the spectrum of fitness values is not sufficient, the fitness function may not provide enough information to guide GP toward regions of the search space, where improved solutions might be found. The fitness function and score were formulated with this in mind and also to reward behaviors responsible for consistently reaching, or coming within close proximity to, the goals.

8.6 *Autonomous navigation results*

In order to demonstrate the operational aspects of the controller in the simplest manner possible, we consider only the composite behavior, **goal-seek**. As illustrated in Figure 8.1, its effect arises from synergistic interaction

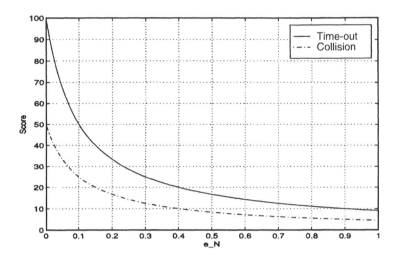

Figure 8.4 Behavior fitness case scoring function.

between primitive behaviors, *go-to-xy* and *avoid-collision*. When more behaviors are involved, the approach proceeds in a straightforward manner by appending additional DOAs and any necessary antecedents to applicability rules accordingly.

The simulated mobile robot is modeled after LOBOt, a custom-built base with a two-wheel differential drive and two stabilizing casters. It is octagonal in shape, 75 cm tall and 60 cm in width. The sensor suite includes optical encoders on each driven wheel and 16 ultrasonic transducers arranged primarily on the front, sides, and forward-facing oblique. Ideal pose information $(x\ y\ \theta)^T$ is assumed and is computed using a kinematic model of the differential-drive mechanism. Its maximum speed was limited to 0.3 m/s which is sufficiently slow to justify the use of a kinematic model only. The sensor model generated range readings with errors as large as ≈ 100 mm and lower resolution than the actual sonar. The simulated operating domain is a hypothetical indoor layout of walls and furniture not unlike an industrial workplace.

8.6.1   Hand-derived behavior

As a result of an arduous trial-and-error procedure, the authors were able to formulate a suitable set of 11 coordination rules for the *goal-seek* behavior. The navigation task of Figure 8.5 illustrates a fragmented view of the primitive capabilities and the result of their modulation. The initial state of the simulation is shown in Figure 8.5(a) with LOBOt located at a docking station with pose $(11.7\ 12.3\ \pi/2)^T$. Its task is to navigate to the goal located at (1.5,1), marked by the X. The *avoid-collision* and *go-to-xy* behaviors are

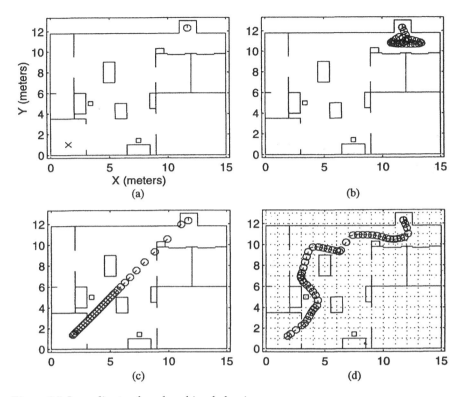

Figure 8.5 Ingredients of goal-seeking behavior.

each shown acting alone in Figure 8.5(b) and (c), respectively. Recall that these behaviors are only capable of exhibiting their particular primitive roles. Thus, *avoid-collision* merely displays cyclic collision-free motion in the immediate vicinity of the robot's initial location, while *go-to-xy* displays a taxic reaction that propels the robot toward the goal regardless of obstacles in its path. Successful completion of the task, resulting from modulation of the primitive behaviors, is shown in Figure 8.5(d). In Figure 8.6, the behavioral interaction during the run is shown as a time history of the DOAs of each primitive behavior. The interaction dynamics show evidence of brief bouts of competition (overlapping oscillations) and cooperation with varying levels of dominance. Initially, *avoid-collision* has the dominant influence over the robot due to the close proximity of walls at the docking station. It virtually maintains dominance throughout the task due to the relatively uniform clutter in the environment. The first bout of competition corresponds to the robot's approach toward the obstacle located at (5,8); a second bout occurs as it enters the goal room. Elsewhere, applicabilities vary continuously, reflecting levels of activation recommended by the behavior control system.

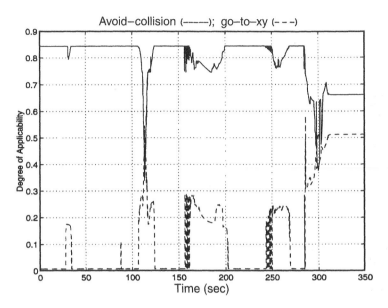

Figure 8.6 Behavior modulation during goal-seeking.

8.6.2   *Evolved behavior*

The operating domain is considerably more general than any of the fitness cases used during the evolution process and, thus, provides a suitable environment to test the generalization capability of the evolved behaviors. The GP system was run using population sizes of 10–20 rule bases for a number of generations ranging from 10–15. In GP, genetic diversity remains high even for very small populations due to the tree structure of individuals (Koza, 1992). Steady-state GP was also applied using a population size of 20. Results of runs using both approaches are summarized graphically in Figure 8.7. The mean performance of GP over five runs is shown, in the left half of Figure 8.7, as the progression of the population average fitness during the first ten generations. The right half of Figure 8.7 shows the progression of the average fitness of the current population at each iteration. Twenty behaviors were processed in the initial population; thereafter, two new behaviors evolved at each iteration. A trend toward higher fitness is evident for both GP and SSGP. Table 8.1 lists some quantitative details about the best behavior evolved by each approach. The success rate was determined from navigation runs in three different simulated domains not included in the set of fitness cases. For this problem, good regions of the search space were discovered with less processing by SSGP.

Having pointed out some operational details of the behavior hierarchy, let us compare the performance of the hand-derived *goal-seek* behavior to a behavior evolved for the same purpose. We will consider an arbitrary point-to-point navigation task from initial state $(1\ 11\text{-}\pi/2)^T$ to a goal

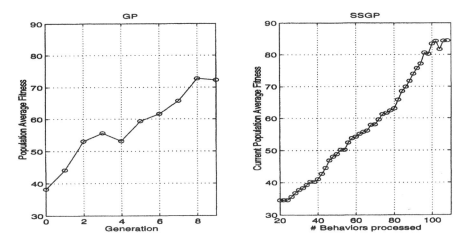

Figure 8.7 Mean performance of GP and SSGP evolution.

Table 8.1		Best Evolved Composite *Goal-Seek* Behaviors			
	Population Size	# Generations	# Rules	Best Fitness	Success Rate (%)
GP	10	15	11	86.5	70
SSGP	20	N/A	9	87.8	83

located at (13.5,4.5). The successful path executed by the hand-derived behavior is shown in Figure 8.8 along with the corresponding behavior modulation history. We compare this with the same task executed by the

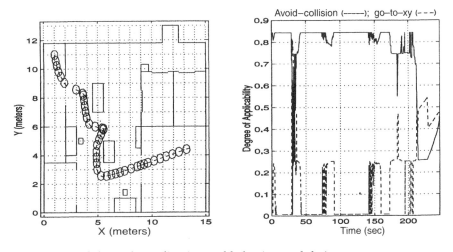

Figure 8.8 Hand-derived coordination and behavior modulation.

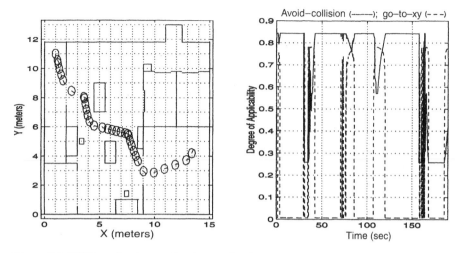

Figure 8.9 SSGP-evolved coordination and behavior modulation.

best SSGP-evolved behavior shown in Figure 8.9. The evolved behavior coordination results in a more direct path to the goal due to higher motivation applied to *go-to-xy*. The resulting path in this example is executed about 20% faster than the path taken via hand-derived coordination. We also note that the behavior modulation demonstrated by the evolved behavior is more complex. Near uniform bouts of competition and cooperation throughout the task are evident in the decision making, thus leading to similar amounts of behavioral influence for each primitive behavior. As listed in Table 8.1, this was achieved using less applicability rules than both the hand-derived behavior and the best GP-evolved behavior. For an identical navigation task, the relative levels of activation induced by the SSGP-evolved goal-seek behavior more closely resemble a consensus.

8.7 Conclusions

The hierarchy of fuzzy behaviors provides an efficient approach to controlling mobile robots. Its practical utility lies in the decomposition of overall behavior into subbehaviors that are activated only when applicable. When conditions for activation of a single behavior (or several) are satisfied, there is no need to process rules from behaviors that do not apply. This would result in unnecessary consumption of computational resources and possible introduction of "noise" into the decision-making process. Behavior modulation allows realization of a continuum of behavioral responses governed by respective behavioral degrees of applicability. This multiple-behavior coordination scheme includes behavior selection as a special case. The modularity and flexibility of the approach, coupled with its mechanisms for weighted decision making, makes it a suitable framework for modeling and controlling situated adaptation in autonomous robots.

Genetic programming proved useful for learning fuzzy behaviors at the coordination level of the hierarchy. In particular, rules were evolved for coordinating low-level fuzzy behaviors that reside at the primitive level. Here, we have focused on the goal-seeking portion of a behavior hierarchy; however, the GP approach is generally applicable throughout. Conventional GP and steady-state GP were applied, each yielding good results for small populations. Overall, SSGP yielded slightly better results for goal-seeking coordination. Using only five fixed fitness cases during behavior evolution, modest generalization capabilities were exhibited by the highest fit behaviors. An SSGP-evolved behavior showed a better behavior modulation capability than both the hand-derived behavior and the best GP-evolved behavior. It is expected that additional improvement can be achieved using a richer set of fitness cases. The results presented here apply only to autonomous navigation in indoor environments. The architecture lecture can be applied for outdoor navigation as well. In future research, implementations for outdoor and rugged terrain vehicles such as planetary rovers will be considered.

Acknowledgments

This work was supported in part by NASA under contracts # NCCW–0087 and Ames-NEG 1198. The author would also like to acknowledge the research assistance of Mr. Harrison Danny and the contributions of Dr. Edward Tunstel, Jr., of Jet Propulsion Laboratory.

References

Beom, H.R., Koh, K.C., and Cho, H.S., Behavioral control in mobile robot navigation using fuzzy decision making approach, *IEEE International Conference on Robots and Systems*, pp. 1938–1945, 1994.

Bonarini, A., Learning to coordinate fuzzy behaviors for autonomous agents, *2nd European Congress on Intelligent Techniques and Soft Computing EUFIT'94*, pp. 475–479, 1994.

Brooks, R.A., A robust layered control system for a mobile robot, *IEEE Journal of Robotics and Automation*, RA-2, 1, 14–23, 1986.

Bruinzeel, J., Jamshidi, M., and Titli, A., A sensory fusion-hierarchical real-time fuzzy control approach for complex systems, Technical Report No. 95347, LAAS-CNRS, Toulouse, France, August 1995.

Correia, L. and Steiger-Gargao, A., A useful autonomous vehicle with a hierarchical behavior control, *Advances in Artificial Life, 3rd European Conference on Artificial Life*, Moran, F. et al., Eds., Springer-Verlag, Granada, Spain, pp. 625–639, June 1995.

Gachet, D. et al., Learning emergent tasks for an autonomous mobile robot, *IEEE International Conference on Robots and Systems*, pp. 290–297, 1994.

Jamshidi, M., Vadiee, N., and Ross, T., Eds., *Fuzzy Logic and Control: Software and Hardware Applications*, Prentice-Hall, Englewood Cliffs, New Jersey, 1993.

Jamshidi, M., Autonomous Control of Complex Systems: Robotic Applications, *Applied Mathmatics and Computations*, 120, pp. 15–29, 2001.

Koza, J.R., *Genetic Programming: On the Programming of Computers by Means of Natural Selection*, MIT Press, Cambridge, Massachusetts, 1992.

Matijevic, J. and Shirley, D., The mission and operation of the Mars Pathfinder Microrover, *IFAC 13th Triennial World Congress*, San Francisco, California, 1996.

Matthias, L. et al., Mars microrover navigation: performance evaluation and enhancement, *Autonomous Robots*, 2, 4, 291–311, 1995.

McFarland, D.J., *Feedback Mechanisms in Animal Behavior*, Academic Press, New York, 1971.

Michaud, F., Lachiver, G., and Le Dinh, C.T., Fuzzy selection and blending of behaviors for situated autonomous agent, *IEEE International Conference on Fuzzy Systems*, pp. 258–264, September 1996.

Moreno, L. et al., Fuzzy supervisor for behavioral control of autonomous systems, *International Conference on Industrial Electronics, Control, and Instrumentation*, pp. 258–261, 1993.

Nordin, P. and Banshaf, W., Genetic programming controlling a miniature robot in real time, Technical Report 4/95, CS Department, University of Dortmund, Germany, 1995.

Pin, F.G. and Watanabe, Y., Navigation of mobile robots using a fuzzy behaviorist approach and custom-designed fuzzy inferencing boards, *Robotica*, 12, 6, 491–504, 1994.

Raju, G.V.S., Zhou, J., and Kisner, R.A., Hierarchical fuzzy control, *International J. of Control*, 54, 5, 1201–1216, 1991.

Reynolds, C.W., Evolution of corridor following behavior in a noisy world, *3rd International Conference on Simulation of Adaptive Behavior*, pp. 402–410, August 1994.

Saffioni, A., Konolige, K., and Ruspini, E.H., A multivalued logic approach to integrating planning and control, *Artificial Intelligence*, 12, 481–526, 1995.

Staddon, J.E.R., *Adaptive Behavior and Learning*, Cambridge University Press, New York, 1983.

Syswerda, G., A study of reproduction in generational and steady-state genetic algorithms, *Foundations of Genetic Algorithms*, Rawlins, G., Ed., Morgan Kaufmann, San Mateo, California, 1991.

Tunstel, E., Coordination of distributed fuzzy behaviors in mobile robot control, *IEEE International Conference on Systems, Man and Cybernetics*, pp. 4009–4014, October 1995.

Tunstel, E. and Jamshidi, M., Fuzzy logic and behavior control strategy for autonomous mobile robot mapping, *IEEE International Conference on Fuzzy Systems*, pp. 514–517, June 1994.

Tunstel, E. and Jamshidi, M., On genetic programming of fuzzy rule-based systems for intelligent control, *Int. Journal of Intelligent Automation and Soft Computing*, 2, 3, 271–284, 1996.

Tunstel, Jr., E. and Jamshidi, M., Intelligent control and evalution of mobile robot behavior, chap. 6 in *Applications of Fuzzy Logic: Towards High MIG Systems*, Jamshidi, M. et al., Eds., Prentice Hall, Saddle River, New Jersey, 1997.

Tunstel, Jr., E., Lippincott, T., and Jamshidi, M., Behavior hierarchy for autonomous mobile robots: Fuzzy-behavior modulation and evolution, *International Journal on Intelligent Automation and Soft Computing*, 3, 1, 37–50, 1997.

chapter nine

Robust control system design: A hybrid H-infinity/multiobjective optimization approach

9.1 Introduction

Recognizing that control system design usually involves the simultaneous consideration of multiple, often competing, performance criteria, these problems can often be formulated as multiobjective optimization (MO) problems. Because the objectives are invariably in competition with each other, there is no unique solution, rather a family of solutions, or *trade-offs*. In the past, multiobjective problems have been cast as, effectively, single objective problems, by constructing an objective function describing the relative importance of each objective. For example, in linear quadratic regulator design, the competing objectives of error and control size have in the past been combined as a weighted sum of quadratic measures. The cost function is defined prior to the optimization procedure. Setting the weights of the cost function requires in-depth information concerning the various trade-offs and valuation of each individual. This data is not commonly fully available in practice.

In recent years, studies have demonstrated that genetic algorithms are a suitable technique for multiobjective optimization, and many examples demonstrate this (e.g., Chipperfield and Fleming, 1996; Dakev et al., 1997; Fonseca and Fleming, 1998b). Due to their population-based nature, GAs are capable of supporting several different solutions simultaneously, and by suitable choice of selection operator, a scheme can be constructed to evolve the trade-off surface, or family of solutions.

In this chapter, the use of *H*-infinity for the design of robust control systems is described (Section 9.2). It is pointed out that the use of the *H*-infinity loop-shaping design procedure requires careful selection of weighting matrices for the satisfaction of performance and robustness specifications. It is suggested that this may be treated as a multiobjective optimization problem. MO is introduced in Section 9.3, and a multiobjective genetic algorithm is described. A benchmark problem, a gasifier system, is introduced in Section 9.4, and it is shown how MO can be used effectively in conjunction with the loop-shaping design procedure to simultaneously produce robust solutions that satisfy performance specifications.

9.2   H-infinity design of robust control systems

9.2.1   Introduction to H-infinity design

H_∞ control system design is a popular optimization-based technique for achieving performance and robustness specifications and is particularly well described in Skogestad and Postlethwaite (1996). It offers the designer a method that can handle multivariable as well as single-loop systems. Levels of robustness can be specified, and trade-offs between performance and robustness in the design procedure can be incorporated. The technique utilizes the related concepts of singular values and the H_∞ norm, the latter being defined for the matrix G:

$$\left\|G(s)\right\|_\infty \quad = \quad \max_\omega \bar{\sigma}(G(j\omega)) \qquad\qquad (9.1)$$

where $\sigma(.)$ denotes the singular values, and $||.||_\infty$ denotes the H_∞ norm.

Classical controller design addresses robustness concerns using gain and phase margin settings. Interactions present in cross-coupled multivariable systems render these methods unreliable as indicators of system robustness. Instead, model uncertainty is incorporated into the design process by representing the plant using a nominal model augmented by a model of the possible uncertainty/perturbation. The controller design strategy is to maximize the size of the modeling error that can be tolerated, while retaining closed-loop stability. In other words, the design procedure seeks to stabilize the set of possible systems that could result from the uncertainty in the plant representation. This is achieved by minimizing the H$_\infty$ norm of the reciprocal of the modeling error.

Performance requirements are addressed through the shaping of frequency responses. This is achieved by using weighting function matrices to provide an acceptable trade-off between disturbance rejection, noise attenuation, and the minimization of control energy. The usual mechanism for doing this is the minimization of the H_∞ norm of the weighted frequency response. The majority of H_∞ techniques address a closed-loop system, and the frequency response used for this exercise is that of the sensitivity function. A popular approach is the H_∞ loop-shaping design procedure of MacFarlane and Glover (1990), and that method is used for the gasifier problem (see Section 9.4, Case Study). The method is unusual in that weighting function matrices are applied to the open-loop plant. This technique has the advantage of offering levels of robust performance, a stronger condition than that of robust stability.

In the H_∞ loop-shaping design procedure, the attainment of performance specifications depends on the selection of the weighting function matrices, denoted W_1 and W_2. It is through this selection process that the designer interacts with the design procedure. This procedure usually involves a degree of trial-and-error style iterative design. Various factors influence the choice of these weighting function matrices, such as bandwidth, roll-off rate, and low-frequency gain magnitude. Certain selection techniques require knowledge of the disturbance process to which the plant is subjected. Redefinition of the performance requirements may be necessary following an unsatisfactory outcome to the design procedure. The designer is faced with a number of considerations that must be balanced against one another in order to achieve the optimum trade-off between performance and robustness. Hence, H_∞ optimization can be viewed as a multiobjective problem.

9.2.2 Loop-shaping design procedure

The H_∞ loop-shaping design procedure is essentially a two-stage controller design technique. First, performance requirements are addressed by shaping the frequency response of the open-loop plant in a manner analogous to that of classical loop shaping. Second, robustness requirements are addressed using H_∞ optimization (Skogestad and Postlethwaite, 1996; Williams, 1991) to stabilize the shaped plant, given a range of possible model uncertainty. The result is a single degree-of-freedom controller; the design procedure assumes positive feedback in the closed-loop system.

In classical linear single-input single-output (SISO) loop shaping, the magnitude of the open-loop transfer function is a function of frequency and is manipulated in order to meet system requirements. The gain of a multi-input multi-output (MIMO) plant, however, varies at any given frequency with the direction of the input vector. No unique gain value can be given for a multivariable system as a function of frequency. A measure analogous to that of SISO plant gain is required for multivariable systems if loop shaping is to be employed. Eigenvalues are unsuitable for this task, as they only provide a measure of gain for the specific case of a square system with input and output vectors that are in the direction of an eigenvector. However, an accurate representation of the gain of a multivariable system can be found using singular value decomposition.

The singular value decomposition of any $l \times m$ matrix G can be written as follows:

$$G = U \Sigma V^H$$

where V^H is the complex conjugate transpose of V. Each column vector of matrix U represents the direction of the vector output signal produced by the plant G subject to an input in the direction of the corresponding column vector of matrix V. These column vectors are each of unit length. Matrix Σ is a diagonal matrix of min$\{l,m\}$ nonnegative singular values in descending order of magnitude, the remaining diagonal elements being zero. These singular values represent the gain of G for the corresponding input and output directions in V and U and can be computed as the positive square roots of the eigenvalues of $G^H G$,

$$\sigma_i(G) = \sqrt{\lambda_i(G^H G)}$$

where $\sigma(.)$ denotes a singular value, and $\lambda(.)$ denotes an eigenvalue.

Hence, the maximum and minimum singular values, $\sigma_{max}(G)$ and $\sigma_{min}(G)$, constitute the upper and lower bounds on the range of system gains in response to all possible input directions at a given frequency.

In order to determine what constitutes a desirable shape for the plant singular values, the closed-loop configuration in Figure 9.1 can be analyzed. From this configuration, the output y can be shown to be

$$y = (I - G_sK)^{-1}G sKr + (I - G_sK)^{-1}G_d d + (I - GsK)^{-1}G_s Kn \qquad (9.2)$$

where r is the reference signal, d is the disturbance, n is the measurement noise, u is the plant input, y is the actual output, and y_m is the measured output.

From Equation (9.2), it can be seen that when $|G_sK|$ is large, reference signals are propagated, while disturbances are attenuated. However, a large value of $|G_sK|$ fails to subdue measurement noise, and a trade-off situation arises. A compromise can be found, because reference signals and disturbances are usually low-frequency events, while measurement noise is prevalent over a much wider bandwidth. Acceptable performance can, therefore, be attained by shaping the singular values of G_sK to give high gain at low

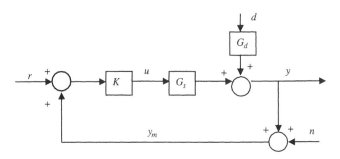

Figure 9.1 One degree of freedom feedback control system. (From Griffin, I.A., Ph.D. thesis, University of Sheffield, U.K., 2002. With permission.)

frequency for disturbance rejection and reduced gain at higher frequency for noise suppression.

For this particular design procedure, the purpose of K is to robustly stabilize the shaped plant as described in the next section; this shaping procedure cannot be accomplished through the manipulation of K. Hence, we define G_s to be the augmented plant,

$$G_s = W_2GW_1 \tag{9.3}$$

where G represents the fixed plant. This structure allows the designer to shape the singular values of the augmented plant G_s, through the selection of appropriate weighting matrices W_1 and W_2. The selection of these matrices is, therefore, the key element in attaining the performance requirements of the system and is the focal point of this design technique. This design task will be performed using a multiobjective GA, which is described in Section 9.3.2.

9.2.3 H-infinity robust stabilization

The normalized left coprime factorization (NLCF) of a plant G is given by $G = M^{-1}N$. A perturbed plant model G_p is then given by the following:

$$G_p = (M + \Delta M)^{-1} (N + \Delta N)$$

To maximize this class of perturbed models such that the configuration shown in Figure 9.2 is stable, a controller K_s that stabilizes the nominal closed-loop system and minimizes γ must be found, where

$$\gamma = \left\| \begin{bmatrix} K_s \\ I \end{bmatrix} (I - GK_s)^{-1} M^{-1} \right\|_\infty$$

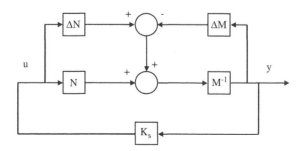

Figure 9.2 Robust stabilization with respect to coprime factor uncertainty. (From Griffin, I.A., Ph.D. thesis, University of Sheffield, U.K., 2002. With permission.)

This is the problem of robust stabilization of normalized coprime factor plant descriptions. From the small gain theorem (Skogestad and Postlethwaite, 1996), the closed-loop plant will remain stable if

$$\left\| \begin{bmatrix} \Delta N \\ \Delta M \end{bmatrix} \right\|_\infty < \gamma^{-1}$$

The lowest possible value of γ and, hence, the highest achievable stability margin is given by $\gamma_{min} = (1 + \rho(ZX))^{\frac{1}{2}}$, where ρ is the spectral radius, and Z and X are the solutions to the following algebraic Riccati equations:

$$(A - BS^{-1}D^TC)Z + Z(A - BS^{-1}D^TC)^T - ZC^TR^{-1}CZ + BS^{-1}B^T = 0 \qquad (9.4)$$

$$(A - BS^{-1}D^TC)^TX + X(A - BS^{-1}D^TC) - XBS^{-1}B^TX + C^TR^{-1}C = 0 \qquad (9.5)$$

where A, B, C, and D are the state space matrices of G and $R = I + DD^T$ and $S = I + D^TD$.

By solving Equations (9.4) and (9.5), the state space controller, K_s, can be generated explicitly (Skogestad and Postlethwaite, 1996). This controller gives no guarantee of the system's performance, simply that it is robustly stable. It is, therefore, necessary to shape the system's response with both pre- and postplant weighting function matrices W_1 and W_2 shown in Figure 9.3. This will ensure that the closed-loop performance meets the specifications required.

Following an introduction to multiobjective optimization (Section 9.3), it will be demonstrated how the design of a robust control system to meet performance specifications may be achieved by using multiobjective optimization to select the parameters of W_1 and W_2 within the framework of an H-infinity design (Section 9.4).

9.3   Multiobjective optimization

9.3.1   Introduction to multiobjective optimization

Multiobjective optimization (MO) recognizes that most practical problems invariably require a number of design criteria to be satisfied simultaneously, viz:

$$\min_{x \in \Omega} \mathbf{f}(\mathbf{x}) \qquad (9.6)$$

where $x = [x_1, x_2, ..., x_q]$, Ω defines the set of free variables, x, subject to any constraints, and $f(x) = [f_1(x), f_2(x), ..., f_n(x)]$ contains the design objectives to be minimized.

Clearly, for this set of functions, $f_i(x)$, it is unlikely that there is one ideal "optimal" solution, rather a set of Pareto-optimal solutions for which an improvement in one of the design objectives will lead to a degradation in

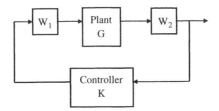

Figure 9.3 Loop-shaping controller structure. (From Griffin, I.A., Ph.D. thesis, University of Sheffield, U.K., 2002. With permission.)

one or more of the remaining objectives. Such solutions are also known as noninferior or nondominated solutions to the MO problem.

The concept of Pareto optimality in the two-objective case is illustrated in Figure 9.4. Here, points A and B are two examples of nondominated solutions on the Pareto front. Neither is preferred to the other. Point A has a smaller value of f_2 than point B, but a larger value of f_1.

9.3.2 *Multiobjective genetic algorithms*

Single-objective optimization has been widely used to address multiobjective optimization problems, but it has a limited capability. Objectives are often noncommensurable and are frequently in conflict with one another. Within a single-objective optimization framework, multiple objectives are often tackled by the "weighted-sum" approach of aggregating objectives. This has a number of significant shortcomings, not least of which is the difficulty of assigning appropriate weights to reflect the relative importance of each objective.

Multiobjective GAs extend standard evolutionary-based optimization techniques to allow individual treatment of several objectives simultaneously. The GA selection operator can be used to identify degrees of Pareto

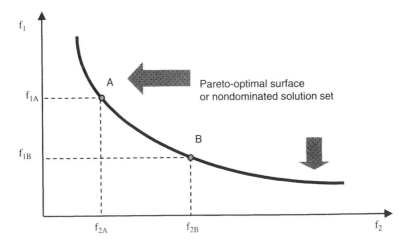

Figure 9.4 Pareto optimality.

optimality, thus enabling objectives to be handled individually. Hence, the requirement for a forced combination of objectives and the need for a priori information are both avoided. Also, the population-based nature of GAs enables them to support several different solutions simultaneously.

One of the first approaches to utilize the concept of Pareto optimality was Fonseca and Fleming's (1993) multiobjective genetic algorithm (MOGA), which tends to be the favored approach for control engineers. Several excellent surveys of multiobjective evolutionary algorithm (MOEA) activity can be found, namely, Veldhuizen and Lamont (2000), Coello (1999), Deb (1999), and Fonseca and Fleming (1995). A book on the subject is also now available (Deb, 2001).

The essential difference between a multiobjective genetic algorithm (MOGA) and a single objective GA is the method by which fitness is assigned to potential solutions. Each solution will have a vector describing its performance across the set of criteria. This vector must be transformed into a scalar fitness value for the purposes of the GA. This process is achieved by ranking the population of solutions relative to each other, and then assigning fitness based on rank. Individual solutions are compared in terms of Pareto dominance. A two-objective minimization example illustration is provided in Figure 9.5. Each member of the population at each generation is examined, and a count is made of those solutions that dominate it. Nondominated solutions are assigned rank 0; those solutions, which are dominated by only one point, are assigned rank 1, and so on. In this way, solutions evolve over the Pareto-optimal surface. In order to encourage an even distribution across this surface, a sharing operator is introduced to inhibit speciation (clustering around regions of the surface); also, a mechanism to discourage the formation of lethals is introduced, suppressing mating pairs from opposite ends of the surface, for example. A pseudocode version of the multiobjective genetic algorithm is provided in Figure 9.6. It has the same basic structure as the genetic algorithm, and the additional features used for handling multiple objectives are highlighted.

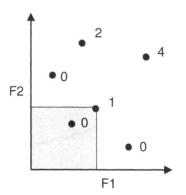

Figure 9.5 Pareto ranking.

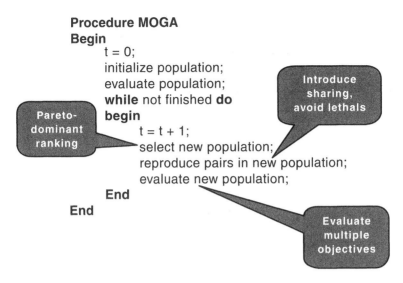

Figure 9.6 Multiobjective genetic algorithm.

For the benchmark problem studied in Section 9.4, an extended feature of MOGA, preference articulation (see Fonseca and Fleming, 1998a), is used. Here, a goal vector, g, is defined, such that the MO problem is now:

$$\min_{x \in \Omega} \mathbf{f}(\mathbf{x})$$

such that

$$\mathbf{f} \leq \mathbf{g} \qquad\qquad (9.7)$$

and the nondominated solution of Equation (9.6) is expected to satisfy Equation (9.7).

9.3.3 *Robust control system design: Incorporating multiobjective with H-infinity*

A multiobjective optimization strategy is compatible with the H_∞ controller design method. The inherent compromise between performance and robustness, which is prevalent in all control system design approaches, lends itself to formulation as a multiobjective H_∞ optimization. The hybrid use of MO and H_∞ is illustrated in Figure 9.7. Within an H_∞ controller design framework, MOGA exercises control over the selection of suitable LSDP weighting matrices to satisfy the performance objectives. These performance objectives are the multiple objectives considered by MOGA. Operating in this way, each controller is assured of being robust, as it has been obtained via the H_∞ design process. The task of the designer will be to

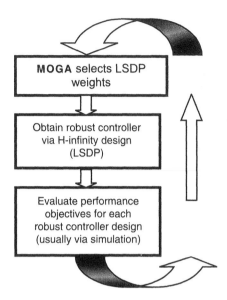

Figure 9.7 Robust control system design: a hybrid MO/*H*-infinity approach.

select a suitable controller from a range of nondominated options. This technique is demonstrated in the next section.

9.4   Case study: Robust control of a gasification plant

The problem studied here arose out of an industrial Benchmark Challenge issued to U.K. Universities in 1997 (Dixon et al., 2000). Modern awareness of environmental issues has led to the desire for low-pollution power generation techniques. A leading member of the Clean Coal Power Generation Group (CCPGG) issued the Challenge. This consortium was established in order to continue the development of a coal-based advanced power generation system that would satisfy the requirement for lower harmful emissions and greater efficiency from a coal-powered electricity generating plant.

In an effort to address this requirement, Integrated Gasification Combined Cycle (IGCC) power plants are being developed around the world. In the United Kingdom, a feasibility study has been conducted on the development of a small-scale Prototype Integrated Plant (PIP) based on the Air Blown Gasification Cycle (ABGC). One of the more advanced components of the PIP, which is not normally encountered in conventional power plants, is the gasification plant (gasifier), the operation of which was based on the spouted fluidized bed concept developed by British Coal. Figure 9.8 contains a functional diagram of the gasifier.

A mixture of ten parts coal to one part limestone is pulverized and conveyed into the gasifier in a stream of air and steam. Once in the gasifier, the air and steam react with the carbon and other volatile elements in the coal. This reaction results in a low calorific value fuel gas and residual ash

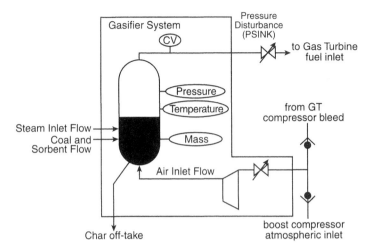

Figure 9.8 Gasifier plant functional diagram. (From Griffin, I.A., Ph.D. thesis, University of Sheffield, U.K., 2002. With permission.)

and limestone derivatives. This residue falls to the bottom of the gasifier and is removed at a controlled rate. The fuel gas escapes through an aperture at the top of the gasifier and is cleaned before being used to power a gas turbine.

9.4.1 Plant model and design requirements

The focus of the Benchmark Challenge was the design of a controller for a single operating point that represented the highly nonlinear gasifier at the 100% load case. Input and output constraints were specified in terms of actuator saturation and rate limits and maximum allowed deviation of controlled variables from the operating point. It was further specified that the controller should be tested at two other linear operating points. These represented the 50 and 0% load cases. This testing was to be performed off-line using the MATLAB®/Simulink analysis and simulation package in order to assess robustness.

The gasifier was modeled as a six-input, four-output multivariable system, and the inputs and outputs of the models are shown in Table 9.1. The state-space representations of the system are linear, continuous, time invariant models and have the inputs and outputs ordered as shown in the table. They represent the system in open-loop and are of 25th order.

The purpose of introducing limestone into the gasifier is to lower the level of harmful emissions. Limestone absorbs sulfur in the coal and, therefore, needs to be introduced at a rate proportional to that of the coal. It was specified in the Challenge that the limestone flow rate should be set to a constant ratio of 1:10, limestone to coal, making the limestone a dependent input. Given that PSINK is a disturbance input, this leaves four degrees of freedom for the controller design, which can, therefore, be treated as a square, 4 × 4 problem.

Table 9.1   Gasifier Inputs and Outputs

Inputs	Outputs
1. WCHR — Char extraction flow (kg/s)	1. CVGAS — Fuel gas calorific value (J/kg)
2. WAIR — Air mass flow (kg)	2. MASS — Bed mass (kg)
3. WCOL — Coal flow (kg/s)	3. PGAS — Fuel gas pressure (N/m²)
4. WSTM — Steam mass flow (kg/s)	4. TGAS — Fuel gas temperature (K)
5. WLS — Limestone mass flow (kg/s)	
6. PSINK — Sink pressure (N/m²)	

The underlying objective of the controller design is to regulate the outputs during disturbance events using the controllable inputs. The outputs of the linear plant models are required to remain within a certain range of the operating point being assessed. These limits are expressed in units relative to the operating point and are therefore the same for all three operating points. They are as follows:

- The fuel gas calorific value (CVGAS) fluctuation should be minimized, but must always be less than +/–10 KJ/kg.
- The bed mass (MASS) fluctuation should be minimized, but must always be less than 5% of the nominal for that operating point.
- The fuel gas pressure (PGAS) fluctuation should be minimized, but must always be less than +/–0.1 bar.
- The fuel gas temperature (TGAS) fluctuation should be minimized, but must always be less than +/–1°C.

The disturbance events stipulated by the Challenge were chosen in order to provide a difficult control design problem. The sink pressure (PSINK) represents the pressure upstream of the gas turbine that the gasifier is ultimately powering. Fluctuations in PSINK represent adjustments to the position of the gas turbine fuel value. The Challenge specified that the value of PSINK should be adjusted in the following ways to generate disturbance events:

- Apply a step change to PSINK of –0.2 bar 30 sec into the simulation. The simulation should be run for a total of 300 sec.
- Apply a sine wave to PSINK of amplitude 0.2 bars and frequency 0.04 Hz for the duration of the simulation. The simulation should be run for a total of 300 sec.

9.4.2   *Problem formulation*

To formulate the MO problem, design objectives, f [Equation (9.6)], are identified along with design goals, g [Equation (9.7)] (because the MOGA

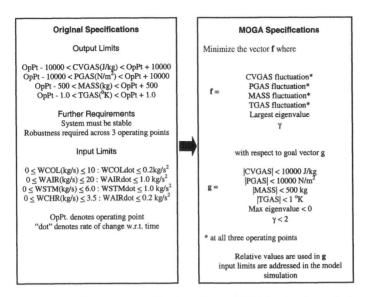

Original Specifications

Output Limits

OpPt - 10000 < CVGAS(J/kg) < OpPt + 10000
OpPt - 10000 < PGAS(N/m²) < OpPt + 10000
OpPt - 500 < MASS(kg) < OpPt + 500
OpPt - 1.0 < TGAS(°K) < OpPt + 1.0

Further Requirements
System must be stable
Robustness required across 3 operating points

Input Limits

$0 \leq$ WCOL(kg/s) ≤ 10 : WCOLdot ≤ 0.2kg/s²
$0 \leq$ WAIR(kg/s) ≤ 20 : WAIRdot ≤ 1.0 kg/s²
$0 \leq$ WSTM(kg/s) ≤ 6.0 : WSTMdot ≤ 1.0 kg/s²
$0 \leq$ WCHR(kg/s) ≤ 3.5 : WAIRdot ≤ 0.2 kg/s²

OpPt. denotes operating point
"dot" denotes rate of change w.r.t. time

MOGA Specifications

Minimize the vector **f** where

$$\mathbf{f} = \begin{matrix} \text{CVGAS fluctuation*} \\ \text{PGAS fluctuation*} \\ \text{MASS fluctuation*} \\ \text{TGAS fluctuation*} \\ \text{Largest eigenvalue} \\ \gamma \end{matrix}$$

with respect to goal vector **g**

$$\mathbf{g} = \begin{matrix} |\text{CVGAS}| < 10000 \text{ J/kg} \\ |\text{PGAS}| < 10000 \text{ N/m}^2 \\ |\text{MASS}| < 500 \text{ kg} \\ |\text{TGAS}| < 1 \text{ °K} \\ \text{Max eigenvalue} < 0 \\ \gamma < 2 \end{matrix}$$

* at all three operating points

Relative values are used in **g**
input limits are addressed in the model
simulation

Figure 9.9 Mapping of system specifications into multiobjective formulation. (From Griffin, I.A., Ph.D. thesis, University of Sheffield, U.K., 2002. With permission.)

preference articulation extension will be used in this study). Figure 9.9 illustrates the conversion of the Benchmark Challenge specifications into MO problem objectives and demonstrates the close mapping between typical control system design objectives and a MO problem formulation amenable to solution by MOGA with the preference articulation extension.

For this problem, the decision variables, x [Equation (9.6)], are the parameters of the elements of the weighting matrices, W_1 and W_2 [Equation (9.3)]. For each set of weighting matrix elements represented by individuals in the populations arising from the MOGA process, the corresponding H_∞ controller is constructed by solving the algebraic Riccati equations (9.4) and (9.5), using the 100% load linear model of the gasifier.

9.4.3 Design using a hybrid H-infinity/multiobjective optimization approach

The weighting function structures used were those of a diagonal matrix of first-order lags for W_1 and a diagonal matrix of gains for W_2. The first-order lag structure of W_1 was considered necessary to break any algebraic loop that may appear in simulation due to the nonzero D matrix in the linear model. As the linear models of the gasifier contain 25 states, this design technique produces controllers that are least of the order of the plant. The terms in W_2 were specified as stateless in order to minimize the order of the resulting controllers. The controller was designed for the 100% load linear model as specified in the Challenge.

Each controller's performance was then evaluated by running simulations at all three operating points using both the step and sine-wave

disturbance signals. As the optimization philosophy of MOGA is to minimize objective function values, the plant model was not off-set, relative
values about the operating point being preferred to absolute input/output
values. This allowed the objective function to assess the peak deviations
in gasifier outputs produced by each candidate controller by taking the
maximum absolute value of each output vector. Input constraints were
observed by placing saturation and rate-limit blocks on the inputs of the
simulated system representation containing relative values appropriate to
the operating point. Stability of the closed-loop system was guaranteed
by minimizing the real part of the closed-loop continuous eigenvalue
having the largest real part, and discarding any individual in the population, which did not result in stable closed-loop eigenvalues. One further
objective attempted to minimize the H_∞ norm, γ [Equation (9.1)], in order
to maximize the robustness of the closed-loop control system.

Figure 9.10 shows a typical "parallel coordinates" trade-off graph for
the gasifier. Trade-off representation between 14 objectives clearly cannot be
represented in the same way as the two-objective case illustrated in Figure
9.4. In this "parallel coordinates representation," each line in the graph
corresponds to a point on the Pareto-optimal curve of Figure 9.4. Each of
these lines connects the performance objectives achieved by an individual
member of the population and represents a potential solution to the design
problem. All solutions illustrated in Figure 9.10 are both nondominant *and*
satisfy the prescribed goals, *g*, as represented by the "X" marks in the figure.
The control system designer, or decision maker (DM), must select a suitable
compromise from this set of solutions. DM may interact with the MOGA as
it evolves, to "tighten" or "slacken" the goals, in order to target a specific
compromise solution.

The 14 objectives shown in Table 9.2 are identified along the x-axis of
Figure 9.10. For example, objective 1 is to minimize the peak fluctuation of
the fuel gas calorific value, CVGAS, from its 100% operating point value.
The cross is situated at the specified limit of 10,000 J/kg. It can clearly be
seen that for objective 1, all solutions have an objective value less than that
of the goal value. This means that all potential controllers offered by the
MOGA have been successful in containing the peak fluctuation of the calorific value at 100% load to within +/−10,000 J/kg.

Figure 9.11 also shows a situation in which all individuals perform
successfully in relation to the performance specifications, but the objectives
have been reordered on the graphical user interface (GUI) to enable comparison between adjacent objectives. The displayed ranges of each objective
are normalized to leave the "X" marks at convenient locations on the graph.
Crossing lines indicate competition between adjacent objectives, whereas
concurrent lines represent noncompeting objectives. Here, the goals relating
to maximum output fluctuation (objectives 1–12) are set to the limits stated
in the Challenge. These targets are specified as constraints in order to guarantee that controllers represented on the trade-off graph satisfy the output
constraints over the run-time of the simulation. From Figure 9.11, it can be

Table 9.2 Design Objectives

Objective Number	Objective Description
1	Peak fluctuation of CVGAS from 100% operating point
2	Peak fluctuation of MASS from 100% operating point
3	Peak fluctuation of PGAS from 100% operating point
4	Peak fluctuation of TGAS from 100% operating point
5	Peak fluctuation of CVGAS from 50% operating point
6	Peak fluctuation of MASS from 50% operating point
7	Peak fluctuation of PGAS from 50% operating point
8	Peak fluctuation of TGAS from 50% operating point
9	Peak fluctuation of CVGAS from 0% operating point
10	Peak fluctuation of MASS from 0% operating point
11	Peak fluctuation of PGAS from 0% operating point
12	Peak fluctuation of TGAS from 0% operating point
13	Maximum continuous eigenvalue of closed-loop system
14	H_∞ robustness measure γ

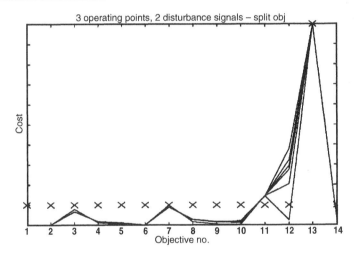

Figure 9.10 Trade-off graph. (From Griffin, I.A., Ph.D. thesis, University of Sheffield, U.K., 2002. With permission.)

seen that all the controllers represented here offer excellent control over peak bed mass fluctuation (objectives 2, 6, and 10). Therefore, the bed mass peak fluctuation, as an objective, is not in competition with any other objective.

The objective visualization tool allows the display of objectives to be reordered, thus providing the user with a means of gaining insight into the nature of conflicts represented by the Pareto-optimal solutions. The peak fluctuation objective values of each output at the three different operating points have been placed adjacent to each other in Figure 9.11. Note that objectives 13 and 14 are not specified in the Benchmark Challenge and are not reordered. This format allows assessment of whether or not the operating points are in competition with each other in terms of each output. The vast majority of solutions are, in fact, concurrent between the objectives relating

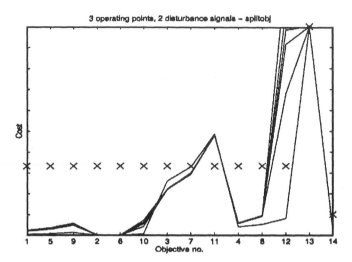

Figure 9.11 Reordered trade-off graph. (From Griffin, I.A., Ph.D. thesis, University of Sheffield, U.K., 2002. With permission.)

to a common output, showing that they are not in competition. The controllers represented will, therefore, optimize the performance at a given output for all operating points. This makes the designer's task easier for this problem, as a final controller selection based on only one of the operating points can be made.

The final choice of controller was made with reference to the performance requirements specified in the Challenge. As can be seen from Figure 9.11, one potential solution satisfied more performance requirements than the others, that being the only one to meet the goal value for objective 12, the peak temperature fluctuation for the 0% operating point. This individual was selected as the final choice of controller. A full set of results and data may be found in Griffin (2002). These results are competitive with, and arguably, superior to those results reported in a special issue (Burnham et al., 2000) devoted to the Challenge. None of the reported controller solutions is able to satisfy all of the design requirements.

The results of the MOGA-based approach are guaranteed to produce a robust solution (because they are the result of an *H*-infinity design) and provide a selection of Pareto-optimal outcomes that are "close" to the prescribed requirements. All that remains for the designer is to select the most appropriate compromise. Griffin (2002) suggests that an acceptable controller might be one that violates only one objective: the peak fluctuation of the gas pressure at the 0% operating point that breaches the constraints specified in the Challenge by less than 50% of the size of the limits. Alternative solutions exist in the Pareto-optimal set, which exerts tighter control over pressure at the expense of calorific value constraint violations. However, these calorific value constraint breaches are extremely large for these controllers, and, as such, these controller choices are unlikely to be suitable.

9.5 Conclusions

This book has investigated methods of designing robust control systems using GAs. In this closing chapter, we further demonstrated the benefits of GAs, this time in providing solutions to multiobjective optimization problems. MOGA has been harnessed effectively with the *H*-infinity loop-shaping design procedure to satisfy particular sets of performance objectives within the overall framework of robust control system design. This has been achieved by using MOGA to manipulate the LSDP weighting matrices in pursuit of the satisfaction of problem requirements.

A moment's reflection will lead one to realize that MOGA could be used in a similar manner for LQG design. Again, for this control system design technique, selection of appropriate integral quadratic weighting matrices, *Q*, *R*, is not straightforward and is often achieved by trial and error. Here, MOGA may be used to assist in weighting matrix parameter selection in pursuit of specific performance objectives, as for the hybrid MOGA/LSDP case (see Figure 9.12). Typically, performance objectives will be evaluated via simulation for each individual arising in the MOGA search process.

These applications of MOGA for control system design fall into the class of "indirect" methods of design. Of course, MOGA may be used in direct approaches, where the decision variables are the controller parameters (see, for example, Fonseca and Fleming, 1998b) or, indeed, a mix of controller structures and parameters, as described in Chipperfield and Fleming 1996).

Multiobjective optimization is a powerful problem formulation representing many real-world problems. Until the benefits of evolutionary computing became apparent, MO problems were, typically, cast as single-

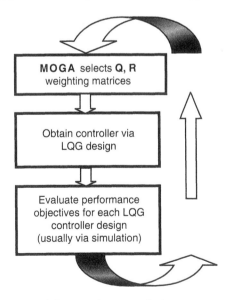

Figure 9.12 MOGA-aided LQG control system design.

objective problems using, for example, a weighted aggregation of objectives, and demanding repeated solution for variants of the chosen single objective, in an attempt to cover the Pareto-optimal surface. The population-based nature of GAs and their flexible search selection operators permit the identification of this Pareto-optimal surface in a *single* run. As a result, evolutionary multicriterion optimization is an active area of study and application, with the potential for wide-ranging benefits, both within control systems engineering and beyond.

References

Burnham, K., Young, P., and Dixon, R., Special Issue on the ALSTOM Gasifier Control Engineering Benchmark Challenge, *Proc. Inst. Mech. Eng., Part I: J. Sys. Contr. Eng.*, 214, 16, 2000.

Chipperfield, A.J. and Fleming, P.J., Multiobjective gas turbine engine controller design using genetic algorithms, *IEEE Trans. Ind. Electron.*, 43, 5, 583–589, 1996.

Coello, C.A.C., A comprehensive survey of evolutionary-based multiobjective optimization techniques, *Knowl. Infor. Sys. Internat. J.*, 1, 3, 269–308, 1999.

Dakev, N.V. et al., Evolutionary H_∞ design of an EMS control system for a Maglev vehicle, *Proc. Inst. Mech. Eng., Part I: J. Sys. Contr. Eng.*, 211, 345–355, 1997.

Deb, K., *Multi-Objective Optimization Using Evolutionary Algorithms*, John Wiley & Sons, Chichester, England, 2001.

Deb, K., Construction of test problems for multi-objective optimization, *GECCO-99: Proceedings of the Genetic and Evolutionary Computation Conference*, 164–171, 1999.

Dixon, R., Pike, A.W., and Donne, M.S., The ALSTOM benchmark challenge on gasifier control, *Proc. Inst. Mech. Eng., Part I: J. Sys. Contr. Eng.*, 214, 389–394, 2000.

Fonseca, C.M. and Fleming, P.J., Genetic algorithms for multiobjective optimisation: Formulation, discussion and generalization, in *Proc. Fifth Inter. Conf. Genetic Algorithms*, San Mateo, California, pp. 416–423, 1993.

Fonseca, C.M. and Fleming, P.J., An overview of evolutionary algorithms in multiobjective optimization, *Evol. Comput.*, 3, 1, 1–16, 1995.

Fonseca, C.M. and Fleming, P.J., Multiobjective optimization and multiple constraint handling with evolutionary algorithms — Part 1: A unified formulation, *IEEE Trans. Sys. Man Cybern.*, 28, 1, 26–37, 1998a.

Fonseca, C.M. and Fleming, P.J., Multiobjective optimization and multiple constraint handling with evolutionary algorithms — Part II: Application example, *IEEE Trans. Sys. Man Cybern.*, 28, 1, 38–47, 1998b.

Griffin, I.A., *Multivariable control methods for gas turbine engines*, Ph.D. dissertation, University of Sheffield, U.K., 2002.

MacFarlane, D.C. and Glover, K., *Robust Controller Design Using Normalised Coprime Factor Plant Descriptions*, Vol. 138, Lecture Notes on Control and Information Science, Springer-Verlag, Berlin, 1990.

Skogestad, S. and Postlethwaite, I., *Multivariable Feedback: Control, Analysis and Design*, John Wiley & Sons, Chichester, England, 1996.

Veldhuizen, D.A.V. and Lamont, G.B., Multiobjective evolutionary algorithms: Analyzing the state-of-the-art, *Evol. Comput.*, 8, 2, 125–147, 2000.

Williams, S.J., H_∞ for the layman, *Meas. Control*, 24, 2, 18–21, 1991.

appendix A

Fuzzy sets, logic and control

A.1 Introduction

One of the more popular new technologies is "intelligent control," which is defined as a combination of control theory, operations research, and artificial intelligence (AI). Judging by the billions of dollars worth of sales and close to 2000 patents issued in Japan alone since the announcement of the first fuzzy chips in 1987, fuzzy logic is still one of the most popular areas in AI.

In order to understand fuzzy logic, it is important to discuss fuzzy sets. In 1965, Zadeh wrote a seminal paper in which he introduced fuzzy sets, i.e., sets with unsharp boundaries. These sets are generally in better agreement with the human mind that works with shades of gray, rather than with just black or white. Fuzzy sets are typically able to represent linguistic terms, e.g., warm, hot, high, low. Today, in Japan, the United States, Europe, Asia, and many other parts of the world, fuzzy technology is widely accepted and applied. Conceptually, a fuzzy set can be defined as a collection of elements in a universe of information, where the boundary of the set contained in the universe is ambiguous, vague, and otherwise fuzzy.

On the other hand, the need and use of multilevel logic can be traced from the ancient works of Aristotle, who is quoted as saying, "There will be a sea battle tomorrow." Such a statement is not yet true or false but is potentially either. Much later, around AD 1285–1340, William of Occam supported two-valued logic but speculated on what the truth value of "if p then q" might be if one of the two components, p or q, was neither true nor false. During the time period of 1878–1956, Lukasiewicz proposed a three-level logic as a "true" (1), a "false" (0), and a "neuter" ($1/2$), which represented half true or half false. In subsequent times, logicians in China and other parts of the world continued on the notion of multilevel logic. Zadeh (1965) finished the task by following through with the speculation of previous logicians and showing that what he called "fuzzy sets" were the foundation of any logic, regardless of the number of truth levels assumed. He chose the innocent word *"fuzzy"* for the continuum of logical values between 0 (completely false) and 1 (completely true). The theory of fuzzy logic deals with two problems: the fuzzy set theory, which deals with the vagueness found in semantics, and the fuzzy measure theory, which deals with the ambiguous nature of judgments and evaluations.

The primary motivation and "banner" of fuzzy logic is the possibility of exploiting tolerance for some inexactness and imprecision. Precision is often costly, so if a problem does not require precision, one should not have to pay for it. The traditional example of parking a car is a noteworthy illustration. If the driver is not required to park the car within an exact distance from the curb, why spend any more time than necessary on the task as long as it is a legal parking operation? Fuzzy logic and classical logic differ in the sense that the former can handle symbolic and numerical manipulation, while the latter can handle symbolic manipulation only. In a broad sense, fuzzy logic is a union of fuzzy (fuzzified) crisp logics (Ross, 1995). To quote Zadeh, "Fuzzy logic's primary aim is to provide a formal, computationally-oriented system of concepts and techniques for dealing with modes of reasoning which are approximate rather than exact." Thus, in fuzzy logic, exact (crisp) reasoning is considered to be the limiting case of approximate reasoning. In fuzzy logic, one can see that everything is a matter of degrees.

A.2 Classical sets

It is instructive to introduce fuzzy sets by first reviewing the elements of classical (crisp) set theory. In classical set theory, a set is denoted as a so-called *crisp set* and can be described by its characteristic function as follows:

$$\mu_C : U \to \{0,1\} \tag{A.1}$$

In the above equation, U is called the universe of discourse, i.e., a collection of elements that can be continuous or discrete. In a crisp set, each element of the universe of discourse either belongs to the crisp set ($\mu_C = 1$) or does not belong to the crisp set ($\mu_C = 0$).

Consider a characteristic function $\mu_{C_{hot}}$ representing the crisp set hot, a set with all "hot" temperatures. Figure A.1 graphically describes this crisp set, considering temperatures higher than 40°C as hot. (Note that for all temperatures T, we have $T \in U$).

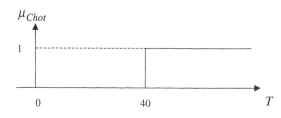

Figure A.1 The characteristic function $\mu_{C_{hot}}$.

A.3 Classical set operations

Let A and B be two sets in the universe U, and $\mu_A(x)$ and $\mu_B(x)$ be the characteristic functions of A and B in the universe of discourse in sets A and B, respectively. The characteristic function $\mu_A(x)$ is defined as follows:

$$\mu_A(x) = \begin{cases} 1, & x \in A \\ 0, & x \notin A \end{cases} \tag{A.2}$$

and $\mu_B(x)$ is defined as

$$\mu_B(x) = \begin{cases} 1, & x \in B \\ 0, & x \notin B \end{cases} \tag{A.3}$$

Using the above definitions, the following operations are defined (Jamshidi et al., 1993):

Union: The union between two sets, i.e., $C = A \cup B$, where $\cup$ is the union operator, represents all those elements in the universe that reside in set A or set B or both (Jamshidi et al., 1993) (see Figure A.2). The characteristic function μ_C is defined below:

$$\forall x \in U : \mu_C = \max\left[\mu_A(x), \mu_B(x)\right] \tag{A.4}$$

The operator in the above equation is referred to as the *max-operator.*
Intersection: The intersection of two sets, i.e., $C = A \cup B$, where $\cap$ is the intersection operator, represents all those elements in the universe U that reside in sets A and B simultaneously (see Figure A.3). Equation (A.5) shows how to obtain the characteristic function μ_C:

$$\forall x \in U : \mu_C = \min\left[\mu_A(x), \mu_B(x)\right] \tag{A.5}$$

The operator in the above equation is referred to as the *min-operator.*
Complement: The complement of a set A, denoted $\overline{A}$, is defined as the collection of all elements in the universe that do not reside in the set A (see Figure A.4). The characteristic function $\mu_{\overline{A}}$ is defined by

$$\forall x \in U : \mu_{\overline{A}} = 1 - \mu_A(x) \tag{A.6}$$

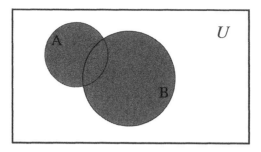

Figure A.2 Union of two sets.

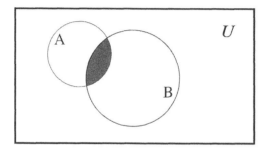

Figure A.3 Intersection of two sets.

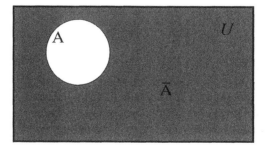

Figure A.4 Complement of set A.

A.4 Properties of classical sets

Properties of classical sets are important to consider because of their influence on mathematical manipulation. Some of these properties are listed below (Ross, 1995):

Commutativity:

$$A \cup B = B \cup A \qquad\qquad (A.7)$$

$$A \cap B = B \cap A \qquad\qquad (A.8)$$

Associativity:

$$A \cup (B \cup C) = (A \cup B) \cup C \qquad \text{(A.9)}$$

$$A \cap (B \cap C) = (A \cap B) \cap C \qquad \text{(A.10)}$$

Distributivity:

$$A \cup (B \cap C) = (A \cup B) \cap (A \cup C) \qquad \text{(A.11)}$$

$$A \cap (B \cup C) = (A \cap B) \cup (A \cap C) \qquad \text{(A.12)}$$

Idempotency:

$$A \cup A = A \qquad \text{(A.13)}$$

$$A \cap A = A \qquad \text{(A.14)}$$

Identity:

$$A \cup \phi = A \qquad \text{(A.15)}$$

$$A \cap X = A \qquad \text{(A.16)}$$

$$A \cap \phi = \phi \qquad \text{(A.17)}$$

$$A \cup X = X \qquad \text{(A.18)}$$

Excluded middle laws are important, because they are the only set operations that are not valid for both classical and fuzzy sets. Excluded middle laws consist of two laws. The first, known as *Law of Excluded Middle*, deals with the union of a set A and its complement. The second law, known as *Law of Contradiction*, represents the intersection of a set A and its complement. The following equations describe these laws:

Law of excluded middle:

$$A \cup \bar{A} = X \qquad \text{(A.19)}$$

Law of contradiction:

$$A \cap \bar{A} = \phi \qquad \text{(A.20)}$$

A.5 Fuzzy sets and membership functions

The definition of a fuzzy set is given by the following characteristic function:

$$\mu_{\tilde{F}}: U \rightarrow [0,1] \qquad \text{(A.21)}$$

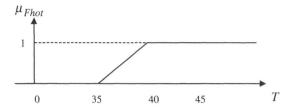

Figure A.5 The membership function μ_{Fhot}.

In this case, the elements of the universe of discourse can belong to the fuzzy set with any value between 0 and 1. This value is called the *degree of membership*. If an element has a value close to 1, the degree of membership, or truth value, is high. The characteristic function of a fuzzy set is called the *membership function*, for it gives the degree of membership for each element of the universe of discourse. If now the characteristic function μ_{Fhot} is considered, one can express the human opinion, for example, that 37°C is still fairly hot, and that 38°C is hot but not as hot as 40°C and higher. This results in a gradual transition from membership (completely true) to nonmembership (not true at all). Figure A.5 shows the membership function μ_{Fhot} for the fuzzy set F_{hot}.

The membership functions for fuzzy sets can have many different shapes, depending on definition. Figure A.6 provides a description of the various features of membership functions. Some of the possible membership functions are shown in Figure A.7.

Figure A.7 illustrates some of the possible membership functions. We have the following: (a) the Γ-function: an increasing membership function with straight lines; (b) the L-function: a decreasing function with straight lines; (c) Λ-function: a triangular function with straight lines; and (d) the

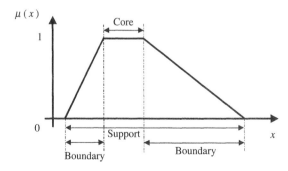

Figure A.6 Description of fuzzy membership functions. (From Jamshidi, M., Vadiee, N., and Ross, T.J., Ed., *Fuzzy Logic and Control: Software and Hardware Applications*, Vol. 2, Prentice Hall Series on Environmental and Intelligent Manufacturing Systems, Jamshidi, M., Ed., Prentice Hall, Englewood Cliffs, New Jersey, 1993. With permission.)

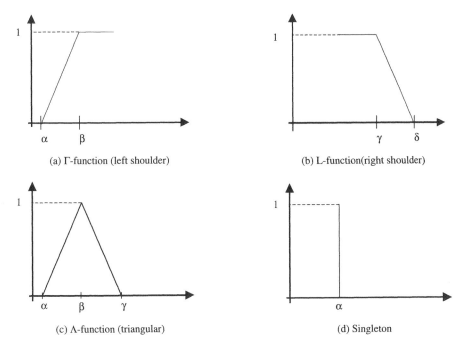

(a) Γ-function (left shoulder)

(b) L-function(right shoulder)

(c) Λ-function (triangular)

(d) Singleton

Figure A.7 Examples of membership functions.

singleton: a membership function with a membership function value 1 for only one value and the rest is zero. There are many other possible functions, such as trapezoidal, Gaussian, sigmoidal, or even arbitrary.

A notation convention for fuzzy sets that is popular in the literature when the universe of discourse U, is discrete and finite, is given below for a fuzzy set A by

$$A = \frac{\mu_{\underset{\sim}{A}}(x_1)}{x_1} + \frac{\mu_{\underset{\sim}{A}}(x_2)}{x_2} + \ldots = \sum_i \frac{\mu_{\underset{\sim}{A}}(x_i)}{x_i} \qquad (A.22)$$

And, when the universe of discourse U is continuous and infinite, the fuzzy set A is denoted by

$$A = \int \frac{\mu_{\underset{\sim}{A}}(x)}{x} \qquad (A.23)$$

A.6 Fuzzy sets operations

As in the traditional crisp sets, logical operations, e.g., union, intersection, and complement, can be applied to fuzzy sets (Jamshidi et al., 1993).

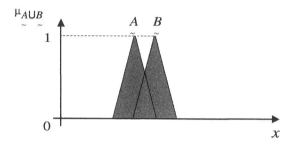

Figure A.8 Union of two fuzzy sets.

Union: The union operation (and the intersection operation as well) can be defined in many different ways. Here, the definition that is used in most cases is discussed (Figure A.8). The union of two fuzzy sets A and B, with the membership functions $\mu_A(x)$ and $\mu_B(x)$, is a fuzzy set C , written as $C = A \cup B$, with a membership function that is related to those of A and B as follows:

$$\forall x \in U : \mu_C = \max\left[\mu_A(x), \mu_B(x)\right] \qquad (A.24)$$

Intersection: According to the *min-operator*, the intersection of two fuzzy sets A and B with the membership functions $\mu_A(x)$ and $\mu_B(x)$, respectively, is a fuzzy set C, written as $C = A \cap B$, with a membership function that is related to those of A and B as follows (Figure A.9):

$$\forall x \in U : \mu_C = \min\left[\mu_A(x), \mu_B(x)\right] \qquad (A.25)$$

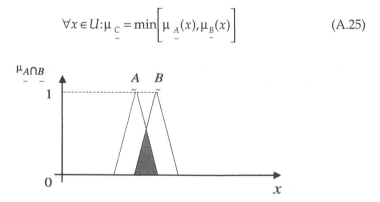

Figure A.9 Intersection of two fuzzy sets.

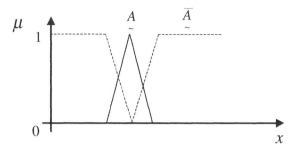

Figure A.10 Complement of a fuzzy set.

Complement: The complement of a set A, denoted $\overline{A}$, is defined as the collection of all elements in the universe that do not reside in the set A (Figure A.10).

$$\forall x \in U : \mu_{\overline{A}} = 1 - \mu_A(x) \tag{A.26}$$

Keep in mind that even though the equations of the union, intersection, and complement appear to be the same for classical and fuzzy sets, they differ in the fact that $\mu_A(x)$ and $\mu_B(x)$ can take only a value of zero or one in the case of classical set, while in fuzzy sets, they include the whole interval from zero to one.

A.7 Properties of fuzzy sets

Similar to classical sets, fuzzy sets also have some properties that are important for mathematical manipulations (Ross, 1995; Dubois and Prade, 1994). Some of these properties are listed below:

Commutativity:

$$A \cup B = B \cup A \tag{A.27}$$

$$A \cap B = B \cap A \tag{A.28}$$

Associativity:

$$A \cup (B \cup C) = (A \cup B) \cup C \tag{A.29}$$

$$A \cap (B \cap C) = (A \cap B) \cap C \tag{A.30}$$

Distributivity:

$$A \cup (B \cap C) = (A \cup B) \cap (A \cup C) \qquad \text{(A.31)}$$

$$A \cap (B \cup C) = (A \cap B) \cup (A \cap C) \qquad \text{(A.32)}$$

Idempotency:

$$A \cup A = A \qquad \text{(A.33)}$$

$$A \cap A = A \qquad \text{(A.34)}$$

Identity:

$$A \cup \phi = A \qquad \text{(A.35)}$$

$$A \cap X = A \qquad \text{(A.36)}$$

$$A \cap \phi = \phi \qquad \text{(A.37)}$$

$$A \cup X = X \qquad \text{(A.38)}$$

Most of the properties that hold for classical sets (e.g., commutativity, associativity, and idempotence) hold also for fuzzy sets, except for the following two properties (Ross, 1995):

Law of contradiction ($A \cap \overline{A} \neq \phi$): One can easily notice that the intersection of a fuzzy set and its complement results in a fuzzy set with membership values of up to $\frac{1}{2}$ and, thus, does not equal the empty set (as in the case of classical sets) as shown in Figure A.11.

Law of excluded middle ($A \cup \overline{A} \neq U$): The union of a fuzzy set and its complement does not give the universe of discourse (see Figure A.12).

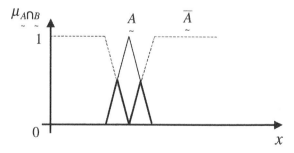

Figure A.11 Law of contradiction.

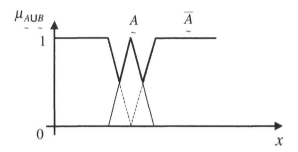

Figure A.12 Law of excluded middle.

Alpha-cut fuzzy sets: It is the crisp domain in which we perform all computations with today's computers. The conversion from fuzzy to crisp sets can be done by two means, one of which is with *alpha-cut sets*.

Given a fuzzy set A , the alpha-cut (or lambda-cut) set of A is defined by

$$A_{\alpha} = \left\{ x \middle| \mu_A(x) \geq \alpha \right\} \qquad (A.39)$$

Note that by virtue of the condition on $\mu_A(x)$ in the above equation, i.e., a common property, the set A_{α} is now a crisp set. In fact, any fuzzy set can be converted to an infinite number of cut sets.

Extension principle: In fuzzy sets, just as in crisp sets, one needs to find means to extend the domain of a function, i.e., given a fuzzy set A and a function $f(\cdot)$, then what is the value of function $f(A)$? This notion is called the extension principle.

Let the function f be defined by

$$f : U \rightarrow V \qquad (A.40)$$

where U and V are domain and range sets, respectively. Define a fuzzy set $A \subset U$ as,

$$A = \left\{ \frac{\mu_1}{u_1} + \frac{\mu_2}{u_2} + \ldots + \frac{\mu_n}{u_n} \right\} \qquad (A.41)$$

Then the extension principle asserts that the function f is a fuzzy set, as well, which is defined below:

$$B = f(A) = \left\{ \frac{\mu_1}{f(u_1)} + \frac{\mu_2}{f(u_2)} + \ldots + \frac{\mu_n}{f(u_n)} \right\} \tag{A.42}$$

The complexity of the extension principle would increase when more than one member of $u_1 \times u_2$ is mapped to only one member of v; one would take the maximum membership grades of these members in the fuzzy set A.

Example A.1

Given two universes of discourse $U_1 = U_2 = \{1,2,\ldots,10\}$ and two fuzzy sets (numbers) defined by

"Approximately 2" $= \dfrac{0.5}{1} + \dfrac{1}{2} + \dfrac{0.8}{3}$ and "Approximately 5" $= \dfrac{0.6}{3} + \dfrac{0.8}{4} + \dfrac{1}{5}$

It is desired to find "approximately 10."

Solution: The function $f = u_1 \_ u_2 :\rightarrow v$ represents the arithmetic product of these two fuzzy numbers and is given by

$$\text{"approximately 10"} = \left(\frac{0.5}{1} + \frac{1}{2} + \frac{0.8}{3} \right) \times \left(\frac{0.6}{3} + \frac{0.8}{4} + \frac{1}{5} \right) = \frac{\min(0.5,0.6)}{3} +$$

$$\frac{\min(0.5,0.8)}{4} + \frac{\min(0.5,1)}{5} + \frac{\min(1,0.6)}{6} + \frac{\min(1,0.8)}{8} +$$

$$\frac{\min(1,1)}{10} + \frac{\min(0.8,0.6)}{9} + \frac{\min(0.8,0.8)}{12} + \frac{\min(0.8,1)}{15} =$$

$$= \frac{0.5}{3} + \frac{0.5}{4} + \frac{0.5}{5} + \frac{0.6}{6} + \frac{0.8}{8} + \frac{0.6}{9} + \frac{1}{10} + \frac{0.8}{12} + \frac{0.8}{15}$$

The above resulting fuzzy number has its *prototype*, i.e., value 10 with a membership function 1, and the other eight pairs are spread around the point (1, 10).

Example A.2

Consider two fuzzy sets (numbers) defined by

"Approximately 2" $= \dfrac{0.5}{1} + \dfrac{1}{2} + \dfrac{0.5}{3}$ and "Approximately 4" $= \dfrac{0.8}{2} + \dfrac{0.9}{3} + \dfrac{1}{4}$

It is desired to find "approximately 8."

Solution: The function $f = u_1 - u_2 :\to v$ represents the arithmetic product of these two fuzzy numbers and is given by

$$"approximately\ 8" = \left(\frac{0.5}{1} + \frac{1}{2} + \frac{0.5}{3}\right) \times \left(\frac{0.8}{2} + \frac{0.9}{3} + \frac{1}{4}\right) = \frac{min(0.5, 0.8)}{2} +$$

$$\frac{min(0.5, 0.9)}{3} + \frac{max[min(0.5, 1), min(1, 0.8)]}{4} +$$

$$\frac{max[min(1, 0.9), min(0.5, 0.8)]}{6} + \frac{min(1, 1)}{8} + \frac{min(0.5, 0.9)}{9} +$$

$$\frac{min(0.5, 1)}{12} = \frac{0.5}{2} + \frac{0.5}{3} + \frac{0.8}{4} + \frac{0.9}{6} + \frac{1}{8} + \frac{0.5}{9} + \frac{0.5}{12}$$

A.8 Predicate logic

Let a predicate logic proposition P be a linguistic statement contained within a universe of propositions that are either completely true or false. The truth value of the proposition P can be assigned a binary truth value, called $T(P)$, just as an element in a universe is assigned a binary quantity to measure its membership in a particular set. For binary (Boolean) predicate logic, $T(P)$ is assigned a value of 1 (truth) or 0 (false). If U is the universe of all propositions, then T is a mapping of these propositions to the binary quantities (0,1), or

$$T:U \to \{0,1\}$$

Now let P and Q be two simple propositions on the same universe of discourse that can be combined using the following five logical connectives — (i) disjunction ($\vee$), (ii) conjunction ($\wedge$), (iii) negation ($-$), (iv) implication ($\to$), and (v) equality ($\leftrightarrow$ or $\equiv$) — to form logical expressions involving two simple propositions. These connectives can be used to form new propositions from simple propositions.

Now define sets A and B from universe X, where these sets might represent linguistic ideas or thoughts. Then, a propositional calculus will exist for the case where proposition P measures the truth of the statement that an element, x, from the universe X is contained in set A, and the truth of the statement that this element, x, is contained in set B, or more conventionally:

P: truth that $x \in A$
Q: truth that $x \in B$, where truth is measured in terms of the truth value,
   i.e.,
If $x \in A$, $T(P) = 1$; otherwise $T(P) = 0$.

If x ∈ B, T(Q) = 1; otherwise T(Q) = 0, or using the characteristic function to represent truth (1) and false (0):

$$\chi_A(x) = \begin{cases} 1, & x \in A \\ 0, & x \notin A \end{cases}$$

The above five logical connectives can be used to create compound propositions, where a compound proposition is defined as a logical proposition formed by logically connecting two or more simple propositions. Just as one is interested in the truth of a simple proposition, predicate logic also involves the assessment of the truth of compound propositions. Given a proposition $P: x \in A, \bar{P}: x \notin A$, the resulting compound propositions are defined below in terms of their binary truth values:

Disjunction:

$$P \vee Q \Rightarrow x \in A \text{ or } B$$

Hence, $T(P \vee Q) = \max(T(P), T(Q))$

Conjunction:

$$P \wedge Q \Rightarrow x \in A \text{ and } B$$

Hence, $T(P \wedge Q) = \min(T(P), T(Q))$

Negation:

If $T(P) = 1$, then $T(\bar{P}) = 0$; If $T(P) = 0$, then $T(\bar{P}) = 1$

Equivalence:

$$P \leftrightarrow Q \Rightarrow x \in A, B$$

Hence, $T(P \leftrightarrow Q) \Rightarrow T(P) = T(Q)$

Implication:

$$P \rightarrow Q \Rightarrow x \notin A \text{ or } x \in B$$

Hence, $T(P \rightarrow Q) = T(\bar{P} \cup Q)$

The logical connective implication presented here is also known as the classical implication, to distinguish it from an alternative form due to

Lukasiewicz, a Polish mathematician in the 1930s, who was first credited with exploring logic other than Aristotelian (classical or binary) logic. This classical form of the implication operation requires some explanation.

For a proposition P defined on set A and a proposition Q defined on set B, the implication "P implies Q" is equivalent to taking the union of elements in the complement of set A with the elements in the set B. That is, the logical implication is analogous to the set-theoretic form.

$$P \rightarrow Q \equiv \overline{A} \cup B \text{ is true } \equiv \text{ either "not in A" or "in B"}$$

So that $(P \rightarrow Q) \leftrightarrow (\overline{P} \vee Q)$

$$T(P \rightarrow Q) = T(\overline{P} \vee Q) = \max(T(\overline{P}), T(Q))$$

This is linguistically equivalent to the statement, "P implies Q is true" when either "*not A*" or "*B*" is true (Jamshidi, 1996). Graphically, this implication and the analogous set operation are represented by the Venn diagram in Figure A.13. As noted, the region represented by the difference $A \backslash B$ is the set region where the implication "P implies Q" is false (the implication *fails*). The shaded region in Figure A.13 represents the collection of elements in the universe, where the implication is true, i.e., the shaded area is the set:

$$\overline{A \backslash B} = \overline{A} \cup B = \overline{(A \cap \overline{B})}$$

If x is in A and x is not in B then

$$A \rightarrow B \equiv \text{ fails } A \backslash B \text{ (difference)}$$

Now, with two propositions (P and Q) each being able to take on one of two truth values (*true* or *false*, 1 or 0), there will be a total of $2^2 = 4$

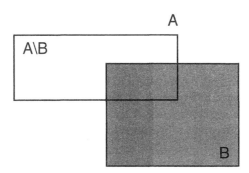

Figure A.13 Classical implication operation (shaded area is where implication holds). (From Ross, T.J., *Fuzzy Logic with Engineering Application*, McGraw-Hill, New York, 1995. With permission.)

Table A.1   Propositional Situations

P	Q	$\overline{P}$	$P \vee Q$	$P \wedge Q$	$P \rightarrow Q$	$P \leftrightarrow Q$
True	True	False	True	True	True	True
True	False	False	True	False	False	False
False	True	True	True	False	True	False
False	False	True	False	False	True	True

propositional situations. These situations are illustrated in Table A.1, along with the appropriate truth values for the propositions P and Q and the various logical connectives between them in the truth table.

To help understand this concept, assume you have two propositions P and Q. P: you are a graduate student, and Q: you are a university student. Let us examine the implication "P implies Q". If you are a student in general, and a graduate student in particular, then the implication is true. On the other hand, the implication would be false if you are a graduate student without being a student. Now, let us assume that you are an undergraduate student; regardless of whether you are a graduate or not, then the implication is true (because in the case you are not a graduate student does not negate the fact that you are an undergraduate). Then, we come to the final case: you are neither a graduate nor undergraduate student. In this case, the implication is true, because the fact that you are not a graduate or undergraduate student does not negate the implication that for you to be a graduate student, you have to be a student at the university.

Suppose the implication operation involves two different universes of discourse, P is a proposition described by set A, which is defined on universe X, and Q is a proposition described by set B, which is defined on universe Y. Then the implication "P implies Q" can be represented in set theory terms by the relation R, where R is defined by

$$R = (A \times B) \cup (\overline{A} \times Y) \equiv \text{IF } A, \text{THEN } B$$

$$\text{If } x \in A \quad (\text{where } x \in X, A \subset X)$$

$$\text{Then } y \in B \quad (\text{where } y \in Y, B \subset Y)$$

where $A \times B$ and $A \times Y$ are Cartesian products (Dubois and Prade, 1980).

This implication is also equivalent to the linguistic rule form: IF A, THEN B. The graphic shown in Figure A.14 represents the Cartesian space of the product $X \times Y$, showing typical sets A and B, and superimposed on this space is the set theory equivalent of the implication. That is,

$$P \rightarrow Q \Rightarrow \text{IF } x \in A, \text{ then } y \in B, \text{ or } P \rightarrow Q \equiv \overline{A} \cup B$$

The shaded regions of the compound Venn diagram in Figure A.14 represent the truth domain of the implication, IF A, THEN B (P implies Q).

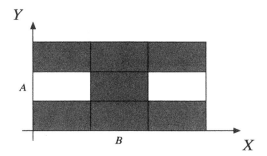

Figure A.14 Cartesian space demonstrating IF *A* THEN *B*.

Tautologies: In predicate logic, it is useful to consider compound propositions that are always true, irrespective of the truth values of the individual simple propositions. Classical logic compound propositions with this property are called *tautologies*. Tautologies are useful for deductive reasoning and for making deductive inferences. So, if a compound proposition can be expressed in the form of a tautology, the truth-value of that compound proposition is known to be true. Inference schemes in expert systems often employ tautologies. The reason for this is that tautologies are logical formulas that are true on logical grounds alone (Dubois and Prade, 1980).

One of these, known as *Modus Ponens* deduction, is a common inference scheme used in forward-chaining rule-based expert systems. It is an operation with the task of finding the truth-value of a consequent in a production rule, given the truth-value of the antecedent in the rule. Modus Ponens deduction concludes that, given two propositions, *a* and *a*-implies-*b*, both of which are true, then the truth of the simple proposition *b* is automatically inferred. Another useful tautology is the *Modus Tollens* inference, which is used in backward-chaining expert systems. In Modus Tollens, an implication between two propositions is combined with a second proposition, and both are used to imply a third proposition. Some common tautologies are listed below.

$$\overline{B} \cup B \leftrightarrow X$$

$$A \cup X \leftrightarrow X$$

$$\overline{A} \cup X \leftrightarrow X$$

$$(A \wedge (A \rightarrow B)) \rightarrow B \qquad (Modus\ Ponens)$$

$$(\overline{B} \wedge (A \rightarrow B)) \rightarrow \overline{A} \qquad (Modus\ Tollens)$$

Contradiction: Compound propositions that are always false, regardless of the truth-value of the individual simple propositions comprising the compound proposition, are called contradictions. Some simple contradictions are listed below.

$$\overline{B} \cap B \leftrightarrow \phi$$

$$A \cap \phi \leftrightarrow \phi$$

$$\overline{A} \cap \phi \leftrightarrow \phi$$

Deductive inferences: The Modus Ponens deduction is used as a tool for inferencing in rule-based systems. A typical IF–THEN rule is used to determine whether an antecedent (cause or action) infers a consequent (effect or action). Suppose we have a rule of the form,

IF A, THEN B

This rule could be translated into a relation using the Cartesian product sets A and B, that is

$$R = (A \times B) \cup (\overline{A} \times Y)$$

Now suppose a new antecedent, say A', is known. Can we use Modus Ponens deduction to infer a new consequent, say B', resulting from the new antecedent? That is, in rule form

IF A', THEN B'?

The answer, of course, is yes, through the use of the composition relation. Because "A implies B" is defined on the Cartesian space $X \times Y$, B' can be found through the following set-theoretic formulation,

$$B' = A' \circ R = A' \circ ((A \times B) \cup (\overline{A} \times Y))$$

Modus Ponens deduction can also be used for the compound rule,

If A, THEN B, ELSE C

Using the relation defined as,

$$R = (A \times B) \cup (\overline{A} \times C)$$

and, hence, $B' = A' \circ R$.

Example A.3

Let two universes of discourse be described by $X = \{1,2,3,4,5,6\}$ and $Y = \{1,2,3,4\}$, and define the crisp set $A = \{2,3\}$ on X and $B = \{3,4\}$ on Y. Determine the deductive inference IF A, THEN B.

Solution: Expressing the crisp sets in fuzzy notation,

$$A = \frac{0}{1} + \frac{1}{2} + \frac{1}{3} + \frac{0}{4}$$

$$B = \frac{0}{1} + \frac{0}{2} + \frac{1}{3} + \frac{1}{4} + \frac{0}{5} + \frac{0}{6}$$

Taking the Cartesian product $A \times B$, which involves taking the pairwise min of each pair from the sets A and B (Jamshidi et al., 1993):

$$A \times B = \begin{array}{c} \\ 1 \\ 2 \\ 3 \\ 4 \end{array} \overset{\displaystyle 1\,2\,3\,4\,5\,6}{\begin{bmatrix} 0\,0\,0\,0\,0\,0 \\ 0\,0\,1\,1\,0\,0 \\ 0\,0\,1\,1\,0\,0 \\ 0\,0\,0\,0\,0\,0 \end{bmatrix}}$$

Then computing $\overline{A} \times Y$,

$$\overline{A} = \frac{1}{1} + \frac{0}{2} + \frac{0}{3} + \frac{1}{4}$$

$$Y = \frac{1}{1} + \frac{1}{2} + \frac{1}{3} + \frac{1}{4} + \frac{1}{5} + \frac{1}{6}$$

$$\overline{A} \times Y = \begin{array}{c} \\ 1 \\ 2 \\ 3 \\ 4 \end{array} \overset{\displaystyle 1\,2\,3\,4\,5\,6}{\begin{bmatrix} 1\,1\,1\,1\,1\,1 \\ 0\,0\,0\,0\,0\,0 \\ 0\,0\,0\,0\,0\,0 \\ 1\,1\,1\,1\,1\,1 \end{bmatrix}}$$

again using pairwise min for the Cartesian product.

The deductive inference yields the following characteristic function in matrix form, following the relation,

$$R = (A \times B) \cup (\bar{A} \times Y) = \begin{matrix} & 1\,2\,3\,4\,5\,6 \\ 1 \\ 2 \\ 3 \\ 4 \end{matrix} \begin{bmatrix} 1\,1\,1\,1\,1\,1 \\ 0\,0\,1\,1\,0\,0 \\ 0\,0\,1\,1\,0\,0 \\ 1\,1\,1\,1\,1\,1 \end{bmatrix}$$

A.9 Fuzzy logic

The extension of the above discussions to fuzzy deductive inference is straightforward. The fuzzy proposition P has a value on the closed interval [0,1]. The truth-value of a proposition $\tilde{P}$ is given by

$$T(\underset{\sim}{P}) = \mu_{\underset{\sim}{A}}(x) \quad \text{where } 0 \leq \mu_{\underset{\sim}{A}} \leq 1$$

Thus, the degree of truth for $P: x \in \underset{\sim}{A}$ is the membership grade of x in A. The logical connectives of negation, disjunction, conjunction, and implication are similarly defined for fuzzy logic, e.g., disjunction.

Negation:

$$T(\underset{\sim}{P}) = \mu_{\underset{\sim}{A}}(x) \quad \text{where } 0 \leq \mu_{\underset{\sim}{A}} \leq 1$$

$$T(\overline{\underset{\sim}{P}}) = 1 - T(\underset{\sim}{P})$$

Disjunction:

$$\underset{\sim}{P} \vee \underset{\sim}{Q} \Rightarrow x \in \underset{\sim}{A} \text{ or } \underset{\sim}{B}$$

$$\text{Hence, } T(\underset{\sim}{P} \vee \underset{\sim}{Q}) = \max(T(\underset{\sim}{P}), T(\underset{\sim}{Q}))$$

Conjunction:

$$\underset{\sim}{P} \wedge \underset{\sim}{Q} \Rightarrow x \in \underset{\sim}{A} \text{ and } \underset{\sim}{B}$$

$$\text{Hence, } T(\underset{\sim}{P} \wedge \underset{\sim}{Q}) = \min(T(\underset{\sim}{P}), T(\underset{\sim}{Q}))$$

Implication:

$$P \rightarrow Q \Rightarrow x \text{ is } A, \text{ then } x \text{ is } B$$

$$T(P \rightarrow Q) = T(\overline{P} \vee Q) = \max(T(\overline{P}), T(Q))$$

Thus, a fuzzy logic implication would result in a fuzzy rule

$$P \rightarrow Q \Rightarrow \text{If } x \text{ is } A, \text{ then } y \text{ is } B$$

and the equivalent to the following fuzzy relation

$$R = (A \times B) \cup (\overline{A} \times Y) \tag{A.1}$$

with a grade membership function,

$$\mu_R = \max \left\{ (\mu_A(x) \wedge \mu_B(y)), (1 - \mu_A(x)) \right\}$$

Example A.4

Consider two universes of discourse described by $X = \{1,2,3,4\}$ and $Y = \{1,2,3,4,5,6\}$. Let two fuzzy sets A and B be given by

$$A = \frac{0.8}{2} + \frac{1}{3} + \frac{0.3}{4}$$

$$B = \frac{0.4}{2} + \frac{1}{3} + \frac{0.6}{4} + \frac{0.2}{5}$$

It is desired to find a fuzzy relation R corresponding to IF A', THEN B'.

Solution: Using the fuzzy relation in Equation A.1 would give

$$
A \underset{\sim}{\times} B =
\begin{array}{c}
\\
1 \\
2 \\
3 \\
4
\end{array}
\begin{array}{cccccc}
1 & 2 & 3 & 4 & 5 & 6 \\
\end{array}
\begin{bmatrix}
0 & 0 & 0 & 0 & 0 & 0 \\
0 & 0.4 & 0.8 & 0.6 & 0.2 & 0 \\
0 & 0.4 & 1 & 0.6 & 0.2 & 0 \\
0 & 0.3 & 0.3 & 0.3 & 0.2 & 0
\end{bmatrix}
$$

$$
\overline{A} \underset{\sim}{\times} Y =
\begin{array}{c}
1 \\
2 \\
3 \\
4
\end{array}
\begin{array}{cccccc}
1 & 2 & 3 & 4 & 5 & 6 \\
\end{array}
\begin{bmatrix}
1 & 1 & 1 & 1 & 1 & 1 \\
0.2 & 0.2 & 0.2 & 0.2 & 0.2 & 0.2 \\
0 & 0 & 0 & 0 & 0 & 0 \\
0.7 & 0.7 & 0.7 & 0.7 & 0.7 & 0.7
\end{bmatrix}
$$

and hence $R = \max\{A \underset{\sim}{\times} B, \overline{A} \underset{\sim}{\times} Y\}$

$$
R =
\begin{array}{c}
1 \\
2 \\
3 \\
4
\end{array}
\begin{array}{cccccc}
1 & 2 & 3 & 4 & 5 & 6 \\
\end{array}
\begin{bmatrix}
1 & 1 & 1 & 1 & 1 & 1 \\
0.2 & 0.4 & 0.8 & 0.6 & 0.2 & 0.2 \\
0 & 0.4 & 1 & 0.6 & 0.2 & 0 \\
0.7 & 0.7 & 0.7 & 0.7 & 0.7 & 0.7
\end{bmatrix}
$$

A.10 Fuzzy control

The aim of this section is to define fuzzy control systems and cover relevant results and development. Traditionally, an *intelligent control* system is defined as one in which classical control theory is combined with artificial intelligence (AI) and possibly OR (Operations Research). Stemming from this definition, two approaches to intelligent control have been in use. One approach combines expert systems in AI with differential equations to create the so-called *expert control*, while the other integrates *discrete event systems* (Markov chains) and differential equations (Wang, 1994). The first approach, although practically useful, is rather difficult to analyze because of the different natures of differential equations (based on mathematical relations) and AI expert systems (based on symbolic manipulations). The second approach, on the other hand, has well-developed and solid theory but is too complex for many practical applications. It is clear, therefore, that a new approach and a change of course are called for here. We begin with another definition of an intelligent control system. An intelligent control system is one in which a physical system or a mathematical model of it is being controlled by a combination of a knowledge-base, approximate (humanlike) reasoning, and a learning process structured in a hierarchical fashion. Under this simple definition, any control system that involves fuzzy logic, neural networks, expert learning schemes, genetic algorithms, genetic programming, or any combination of these would be designated as intelligent control.

Among the many applications of fuzzy sets and fuzzy logic, fuzzy control is perhaps the most common. Most industrial fuzzy logic applications in Japan, the United States, and Europe fall under fuzzy control. The reasons for the success of fuzzy control are both theoretical and practical (Wang, 1994; Jamshidi, 1996).

From a theoretical point of view, a fuzzy logic rule base can be used to identify a model, as a "universal approximation," as well as a nonlinear controller. The most relevant information about any system comes in one of three ways — a mathematical model, sensory input/output data, and human expert knowledge. The common factor in all three sources is knowledge. For many years, classical control designers began their effort with a mathematical model and did not go any further in acquiring more knowledge about the system, i.e., designers put their entire trust in a mathematical model with accuracy that may sometimes be in question. Today, control engineers can use all of the above sources of information. Aside from a mathematical model with utilization that is clear, numerical (input/output) data can be used to develop an approximate model (input/output nonlinear mapping) as well as a controller, based on the acquired fuzzy IF-THEN rules.

Some researchers and teachers of fuzzy control systems subscribe to the notion that fuzzy controls should always use a model-free design approach and, hence, give the impression that a mathematical model is irrelevant. As indicated before, the authors, however, believe strongly that if a mathematical model exists, it would be the first source of knowledge used in building the entire knowledge base. From a mathematical model, through simulation, for example, one can further build the knowledge base. Through utilization of the expert operator's knowledge which comes in the form of a set of linguistic or semilinguistic IF-THEN rules, the fuzzy controller designer would get a big advantage in using every bit of information about the system during the design process.

On the other hand, it is quite possible that a system, such as high-dimensional large-scale systems, is so complex that a reliable mathematical tool either does not exist or is costly to attain. This is where fuzzy control (or intelligent control) comes in. Fuzzy control approaches these problems through a set of local humanistic (expert-like) controllers governed by linguistic fuzzy IF-THEN rules. In short, fuzzy control falls into the category of intelligent controllers, which are not solely model based but are also knowledge based.

From a practical point of view, fuzzy controllers, which have appeared in industry and in manufactured consumer products, are easy to understand, simple to implement, and inexpensive to develop. Because fuzzy controllers emulate human control strategies, they are easily understood even by those who have no formal background in control. These controllers are also simple to implement.

A.11   Basic definitions

A common definition of a fuzzy control system is that it is a system that
emulates a human expert. In this situation, the knowledge of the human
operator would be put in the form of a set of fuzzy linguistic rules. These
rules would produce an approximate decision, just as a human would.
Consider Figure A.15, where a block diagram of this definition is shown. As
shown, the human operator observes quantities by observing the inputs, i.e.,
reading a meter or measuring a chart, and performs a definite action (e.g.,
pushes a knob, turns on a switch, closes a gate, or replaces a fuse), thus
leading to a crisp action, shown here by the output variable $y(t)$. The human
operator can be replaced by a combination of a fuzzy rule-based system
(FRBS) and a block called *defuzzifier*. The input sensory (crisp or numerical)
data are fed into FRBS, where physical quantities are represented or compressed into linguistic variables with appropriate membership functions.
These linguistic variables are then used in the *antecedents* (IF-Part) of a set
of fuzzy rules within an inference engine to result in a new set of fuzzy
linguistic variables or *consequent* (THEN-Part). Variables are then denoted in
this figure by z and are combined and changed to a crisp (numerical) output
$y^*(t)$, which represents an approximation to actual output $y(t)$.

It is, therefore, noted that a fuzzy controller consists of three operations:
(1) fuzzification, (2) inference engine, and (3) defuzzification.

Before a formal description of the fuzzification and defuzzification processes is made, let us consider a typical structure of a fuzzy control system,
which is presented in Figure A.16. As shown, the sensory data go through
two levels of interface, i.e., the analog to digital and the crisp to fuzzy and,
at the other end, in reverse order, i.e., fuzzy to crisp and digital to analog.

Another structure for a fuzzy control system is a fuzzy inference, connected to a knowledge base, in a supervisory or adaptive mode. The structure
is shown in Figure A.17. As shown, a classical crisp controller (often an
existing one) is left unchanged, but through a fuzzy inference engine or a
fuzzy adaptation algorithm, the crisp controller is altered to cope with the
system's unmodeled dynamics, disturbances, or plant parameter changes,

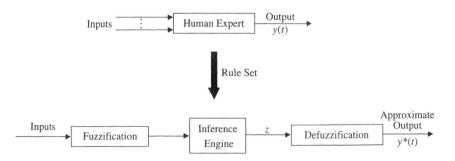

Figure A.15 Conceptual definition of a fuzzy control system.

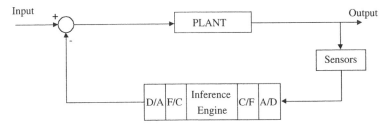

Figure A.16 Block diagram for a laboratory implementation of a fuzzy controller.

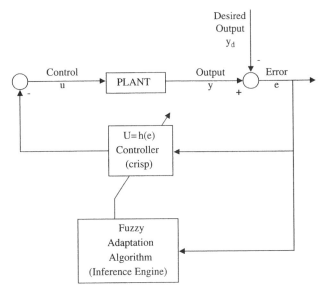

Figure A.17 An adaptive (tuner) fuzzy control system.

much like a standard adaptive control system. Here, the function $h(\cdot)$ represents the unknown nonlinear controller or mapping function $h: e \rightarrow u$, which along with any two input components e_1 and e_2 of e, represents a nonlinear surface, sometimes known as the *control surface* (Jamshidi, 1996).

The fuzzification operation, or the *fuzzifier* unit, represents a mapping from a crisp point $x = (x_1\, x_2\, \dots\, x_n)^T \in X$ into a fuzzy set $A \in X$, where X is the universe of discourse and T denotes vector or matrix transposition.* There are normally two categories of fuzzifiers in use. The first is singleton, and the second is nonsingleton. A singleton fuzzifier has one point (value) x_p as its fuzzy set support, i.e., the membership function is governed by the following relation:

---

* For convenience, in this section, the tilde ($\sim$) sign that was used earlier to express fuzzy sets is omitted.

$$\mu_A(x) = \begin{cases} 1, & x = x_p \in X \\ 0, & x \neq x_p \in X \end{cases} \qquad (A.43)$$

The nonsingleton fuzzifiers are those in which the support is more than a point. Examples of these fuzzifiers are triangular, trapezoidal, Gaussian, etc. In these fuzzifiers, $\mu_A(x) = 1$ at $x = x_p$ at $x=xp$, where xp may be one or more than one point, and then $\mu_A(x)$ decreases from 1 as x moves away from x_p or the "core" region to which x_p belongs such that $\mu_A(x_p)$ remains 1 (see Section A.5). For example, the following relation represents a Gaussian-type fuzzifier:

$$\mu_A(x) = \exp\left\{ -\frac{(x-x_p)^T(x-x_p)}{\sigma^2} \right\} \qquad (A.44)$$

where the variance, σ^2, is a parameter characterizing the shape of $\mu_A(x)$.

Inference engine: The cornerstone of any expert controller is its inference engine, which consists of a set of expert rules, which reflect the knowledge base and reasoning structure of the solution of any problem. A fuzzy (expert) control system is no exception, and its rule base is the heart of the nonlinear fuzzy controller. A typical fuzzy rule can be composed as (Jamshidi et al., 1993):

$$\text{IF } A \text{ is } A_1 \text{ AND } B \text{ is } B_1 \text{ OR } C \text{ is } C_1 \text{ THEN } U \text{ is } U_1 \qquad (A.45)$$

where A, B, C, and U are fuzzy variables, A_1, B_1, C_1, and U_1 are fuzzy linguistic values (membership functions or fuzzy linguistic labels), "AND," "OR," and "NOT" are connectives of the rule. The rule in Equation A.45 has three antecedents and one consequent. Typical fuzzy variables may, in fact, represent physical or system quantities such as: "temperature," "pressure," "output," "elevation," etc., and typical fuzzy linguistic values (labels) may be "hot," "very high," "low," etc. The portion "very" in a label "very high" is called a *linquistic hedge*. Other examples of a hedge are "much," "slightly," "more," or "less," etc. The above rule is known as the Mamdani-type rule. In Mamdani rules, the antecedents and the consequent parts of the rule are expressed using linguistic labels. In general in fuzzy system theory, there are many forms and variations of fuzzy rules, some of which will be introduced here and throughout the section. Another form is *Takagi–Sugeno* rules, in which the consequent part is expressed as an analytical expression or equation.

Two cases will be used here to illustrate the process of inferencing graphically. In the first case, the inputs to the system are crisp values, and we use max-min inference method. In the second case, the inputs to the system are also crisp, but we use the max-product inference method. Please keep in mind that there could also be cases where the inputs are fuzzy variables.

Consider the following rule with a consequent that is not a fuzzy implication:

$$\text{IF } x_1 \text{ is } A_1^i \text{ AND } x_2 \text{ is } A_2^i \text{ THEN } y^i \text{ is } B^i, \text{ for } i = 1, 2, ..., l \qquad (A.46)$$

where A_1^i and A_2^i are the fuzzy sets representing the ith-antecedent pairs, and B^i are the fuzzy sets representing the ith-consequent, and l is the number of rules.

Case A.1

Inputs x_1 and x_2 are crisp values, and max-min inference method is used. Based on the Mamdani implication method of inference, and for a set of *disjunctive rules*, i.e, rules connected by the *OR* connective, the aggregated output for the l rules presented in Equation A.46 will be given by

$$\mu_{B^i}(y) = \max_i[\min[\mu_{A_1^i}(x_1), \mu_{A_2^i}(x_2)]], \text{ for } i = 1, 2, ..., l \qquad (A.47)$$

Figure A.18 is a graphical illustration of Equation A.47, for $l = 2$, where A_1^1 and A_2^1 refer to the first and second fuzzy antecedents of the first rule, respectively, and B^1 refers to the fuzzy consequent of the first rule. Similarly, A_1^2 and A_2^2 refer to the first and second fuzzy antecedents of the second rule, respectively, and B^2 refers to the fuzzy consequent of the second rule. Because the antecedent pairs used in general form presented in Equation A.46 are connected by a logical *AND*, the minimum function is used. For each rule, minimum value of the antecedent propagates through and truncates the membership function for the consequent. This is done graphically for each rule. Assuming that the rules are disjunctive, the aggregation operation *max* results in an aggregated membership function comprised of the outer envelope of the individual truncated membership

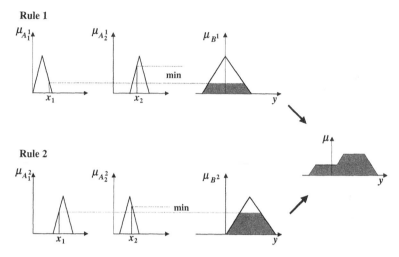

Figure A.18 Graphical representation of Max-Min inference rules.

forms from each rule. To compute the final crisp value of the aggregated output, defuzzification is used, which will be explained in the next section.

Case A.2

Inputs x_1 and x_2 are crisp values, and max-product inference method is used. Based on the Mamdani implication method of inference, and for a set of *disjunctive rules*, the aggregated output for the l rules presented in Equation A.47 will be given by

$$\mu_{B^i}(y) = \max_i [\mu_{A_1^i}(x_1) \cdot \mu_{A_2^i}(x_2)], \text{ for } i = 1, 2, ..., l \qquad (A.48)$$

Figure A.19 is a graphical illustration of Equation A.48, for $l = 2$, where A_1^1 and A_2^1 refer to the first and second fuzzy antecedents of the first rule, respectively, and B^1 refers to the fuzzy consequent of the first rule. Similarly, A_1^2 and A_2^2 refer to the first and second fuzzy antecedents of the second rule, respectively, and B^2 refers to the fuzzy consequent of the second rule. Because the antecedent pairs used in general form presented in Equation A.47 are connected by a logical *AND*, the minimum function is used again. For each rule, minimum value of the antecedent propagates through and scales the membership function for the consequent. This is done graphically for each rule. Sim-

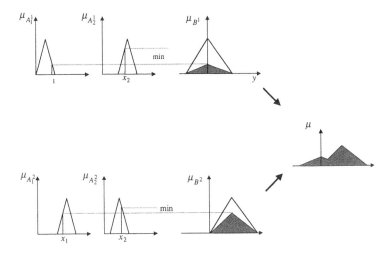

Figure A.19 Graphical representation of Max-Product inference rules.

ilar to the first case, the aggregation operation *max* results in an aggregated membership function comprised of the outer envelope of the individual truncated membership forms from each rule. To compute the final crisp value of the aggregated output, defuzzification is used.

Defuzzification: Defuzzification is the third important element of any fuzzy controller. In this section, only the *center of gravity defuzzifier*, which is the most common one, is discussed. In this method, the weighted average of the membership function or the center of gravity of the area bounded by the membership function curve is computed as the most typical crisp value of the union of all output fuzzy sets:

$$y_c = \frac{\int y \cdot \mu_A(y)dy}{\int \mu_A(y)dy} \tag{A.49}$$

Fuzzy control design: One of the first steps in the design of any fuzzy controller is to develop a knowledge base for the system to eventually lead to an initial set of rules. There are at least five different methods to generate a fuzzy rule base (Jamshidi, 1996):

Simulate the closed-loop system through its mathematical model.
Interview an operator who has had many years of experience controlling the system.
Generate rules through an algorithm using numerical input/output data of the system.

Use learning or optimization methods such as neural networks (NN) or genetic algorithms (GA) to create the rules.

In the absence of all of the above, if a system exists, experiment with it in the laboratory or factory setting and gradually gain enough experience to create the initial set of rules.

Example A.5

Consider the linearized model of the inverted pendulum (Figure A.20), described by the equation given below (Zilouchian and Jamshidi, 2001),

$$\dot{x} = \begin{pmatrix} 0 & 1 \\ 15.79 & 0 \end{pmatrix} x + \begin{pmatrix} 0 \\ 1.46 \end{pmatrix} u$$

with $l = 0.5$ m, $m = 100$ g, and initial conditions $x^T(0) = [\theta(0) \ \dot{\theta}(0)]^T = \dot{\theta}[1 \ 0]^T$. It is desired to stabilize the system using fuzzy rules.

Solution: Clearly, this system is unstable, and a controller is needed to stabilize it. To generate the rules for this problem, only common sense is needed, i.e., if the pole is falling in one direction, then push the cart in the same direction to counter the movement of the pole. To put this into rules of the form of Equation A.47 we get the following:

IF θ is θ_Positive AND $\dot{\theta}$ is $\dot{\theta}$_Positive THEN u is u_Negative

$$(A.50)$$

IF θ is θ_Negative AND $\dot{\theta}$ is $\dot{\theta}$_Negative THEN u is u_Positive

where the membership functions described above are defined in Figure A.21.

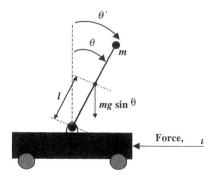

Figure A.20 The inverted pendulum problem.

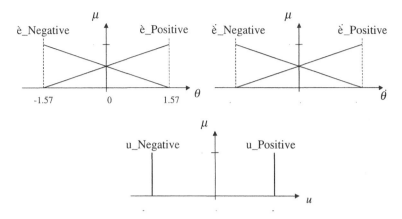

Figure A.21 Membership functions for the inverted pendulum problem.

As shown in Figure A.21, the membership functions for the inputs are half-triangular, while the membership function of the output is singleton. By simulating the system with fuzzy controller, we get the response shown in Figure A.22. It is clear that the system is stable. In this example, only two rules were used, but more rules could be added in order to get a better response, i.e., less undershoot.

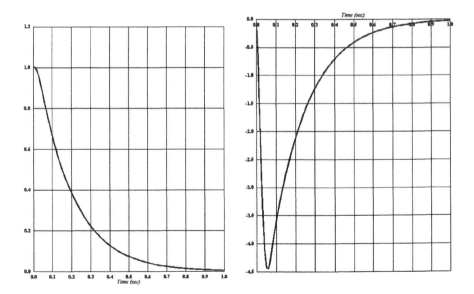

Figure A.22 Simulation results for Example A.5.

A.12   Conclusion

In this appendix, a quick overview of classical and fuzzy sets, classical and fuzzy logic, and fuzzy control were given. Main similarities and differences between classical and fuzzy sets were introduced. In general, set operations are the same for classical and fuzzy sets. The exceptions were excluded middle laws. Alpha-cut sets and extension principle were presented followed by a brief introduction to classical versus fuzzy relations. This section presented issues that are important in understanding fuzzy sets and their advantages over classical sets. Most of the tools needed to form an idea about fuzzy logic and its operation have been introduced. These tools are essential in understanding fuzzy control and fuzzy-GA control approaches in the text. Fuzzy control systems are desirable in situations where precise mathematical models are not available and human involvement is necessary. In that case, fuzzy rules could be used to mimic human behavior and actions.

References

Dubois, D. and Prade, H., *Fuzzy Sets and Systems, Theory and Applications*, Academic Press, New York, 1994.

Jamshidi, M., *Large-Scale Systems — Modeling, Control and Fuzzy Logic*, Prentice Hall Series on Environmental and Intelligent Manufacturing Systems, Jamshidi, M., Ed., Vol. 8, Saddle River, New Jersey, 1996.

Jamshidi, M., Vadiee, N., and Ross, T.J., Ed., *Fuzzy Logic and Control: Software and Hardware Applications*, Vol. 2, Prentice Hall Series on Environmental and Intelligent Manufacturing Systems, Jamshidi, M., Ed., Prentice Hall, Englewood Cliffs, New Jersey, 1993.

Ross, T.J., *Fuzzy Logic with Engineering Application*, McGraw-Hill, New York, 1995.

Wang, L.-X., *Adaptive Fuzzy Systems and Control*, Prentice Hall, Englewood Cliffs, New Jersey, 1994.

Zadeh, L.A., Fuzzy sets, *Inf. C.*, 8, 338–353, 1965.

Zilouchian, A. and Jamshidi, M., *Intelligent Control Systems Using Soft Computing Methodologies*, CRC Press, Boca Raton, Florida, 2001.

Index

Milton Keynes UK
Ingram Content Group UK Ltd.
UKHW020313111024
449327UK00040B/705